Renewable Energy
for Residential Heating
and Cooling

Marbek

Marbek has been helping clients in Canada and around the world find solutions to energy and environmental challenges since 1983, completing over $40 million of work and 1200 projects in more than 30 countries. We offer multidisciplinary expertise and experience in the public, private and voluntary sectors, specializing in energy and environmental policy, programmes, technical services and management consulting. We help clients chart a clearer path toward positive organizational, social and environmental change, and we do it with pride, integrity and an uncompromising commitment to quality.

WIP Renewable Energies

WIP Renewable Energies has been active in the fields of renewable energy and biofuel technologies for more than three decades, providing a range of technical expertise and non-technical services to both industrial and public sector clients at the international level. WIP has a long record of performing outstanding activities for and in cooperation with the European Commission and German national entities. These activities include: analytical research on energy issues; renewable energy (RE) systems; energy projects in European Union (EU) countries; strategies for greater integration of RE technologies in EU and emerging countries (such as Latin America, Africa, China); event management for European conferences, workshops, campaigns and expert services in international RE policy and research. WIP is member of the European Biomass, Photovoltaic and Wind Energy Associations (EUBIA, EPIA, EWEA), as well as a founding member of the European Renewable Energy Research Centres (EUREC) Agency.

IEA-RETD

The IEA Implementing Agreement on Renewable Energy Technology Deployment (IEA-RETD) was formally established in 2005. The mission of the IEA-RETD (www.iea-retd.org) is to act as a catalyst for an increased rate of renewable energy technologies deployment by:

- Providing evidence about the value and benefits of renewable energy.
- Facilitating and fostering ongoing international dialogue.
- Creating better awareness of the importance and the potential of renewable energy deployment.
- Networking to improve the dialogue between the private and public sector.
- Strengthening communication and dialogue within the IEA network.

Renewable Energy for Residential Heating and Cooling

Policy Handbook

Edited by
IEA-RETD

London • Washington, DC

First published in 2011 by Earthscan

Copyright ©IEA-RETD, 2011

Earthscan Ltd, Dunstan House, 14a St Cross Street, London EC1N 8XA, UK
Earthscan LLC, 1616 P Street, NW, Washington, DC 20036, USA
Earthscan publishes in association with the International Institute for Environment and Development

For more information on Earthscan publications, see www.earthscan.co.uk or write to earthinfo@earthscan.co.uk

ISBN: 978-1-84971-278-1 hardback

Typeset by Saxon Graphics Ltd, Derby
Cover design by Susanne Harris

A catalogue record for this book is available from the British Library

Library of Congress Cataloging-in-Publication Data
Renewable energy for residential heating and cooling policy handbook / edited by IEA-RETD.
     p. cm.
  Includes bibliographical references and index.
  ISBN 978-1-84971-278-1 (hardback)
  1. Dwellings—Heating and ventilation—Handbooks, manuals, etc. 2. Dwellings—Air conditioning—Handbooks, manuals, etc. 3. Renewable energy sources—Handbooks, manuals, etc. I. IEA-RETD (Organization)
  TH7222.R46 2011
  333.79'63—dc22   2010035504

Printed and bound in the UK by TJ International Ltd.
The paper used is FSC certified and the inks are vegetable based.

Contents

List of Exhibits

Part 2 Best Practices Guide

Acknowledgements

The authors would like to thank the project steering committee for project guidance and review. Committee members include Michael Paunescu (Chair, Natural Resources Canada), Milou Beerepoot (Renewable Energy Division of IEA, France), Even Bjørnstad (Enova SF, Norway) and Kristian Petrick (Operating Agent, Factor CO2 Ennova, Spain). Lex Bosselaar (Agentschap NL, Executive Committee Member of the IEA Solar Heating and Cooling Programme) also provided project review.

The authors would also like to thank the programme administrators interviewed during the best practices investigation. The generosity of these individuals greatly enhanced the quality of the information in this report.

Executive Summary

Success in the development and use of renewable energy for heating and cooling (REHC) technologies has varied by continent, country and even by municipality. Not surprisingly, some regions with rapidly developing REHC markets have abundant renewable resources or unreliable or expensive conventional fuel sources. The success of other leading regions cannot be explained by external factors; it appears this success is due to successful REHC promotion programmes. This book details a project designed to extract best practices and lessons learned from successful REHC programmes for transfer to other regions that have been, to this point, less successful in developing REHC markets.

The 2007 joint International Energy Agency (IEA) and the IEA-RETD report *Renewables for Heating and Cooling – Untapped Potential*[1] recommended that a review be conducted of best practices and lessons learned from programmes supporting REHC technologies in leading countries. Moreover, the report recommended a separate examination of these practices for the residential sector.

In response, this current project was commissioned under the Implementing Agreement on Renewable Energy Technology Deployment (IEA-RETD).[2] Its goals were to:

- Examine the REHC technology status in IEA-RETD member countries and others.
- Identify and investigate significant residential-sector REHC support programmes in these countries.
- Extract and analyse a set of best practices from these programmes.

The overall objective of this project was to develop materials 'to assist policymakers in developing robust, effective, sustainable programmes that can overcome the most common barriers that occur in deploying REHC in the residential sector'.[3]

The end products of this project are two documents: Part 1 of this book (Best Practices in the Deployment of Renewable Energy for Heating and Cooling in the Residential Sector) and Part 2 (Best Practices Guide), which is intended for programme developers at all jurisdictional levels.

Project scope

The scope of this project was designed to develop a body of information broad enough to be useful in a variety of circumstance, but deep enough to be meaningful. The scope covers the full range of situations defined by:

- Programme phases: portfolio planning, programme design, programme implementation and evaluation.
- Market maturity stages: initial deployment, mass market and full market.
- Instrument categories: economic incentives, regulations, and information and market activities.
- Applicable residential sectors: new and existing buildings, including both single and multi-family dwellings.
- Four categories of technologies:
  - active solar thermal for air and water heating;
  - biomass (pellets, wood and wood waste);
  - geothermal (ground-source heat pumps);
  - heat pump technologies based on ambient air heat (air-to-air and air-to-liquid).

The intended readership for the Best Practices Guide includes policymakers as well as programme designers, implementers and evaluators at all government levels and within other agencies or utilities involved in encouraging the deployment of REHC technologies.

Methodology

As a first step in identifying successful programmes, country profiles were developed for the ten IEA-RETD member countries: Canada, Denmark, France, Germany, Ireland, Italy, Japan, the Netherlands, Norway and the UK. In addition, Austria and Spain were also profiled due to their success with renewable energy; the US was included due to both programme success and variety; and an abbreviated summary of the rapidly developing solar thermal energy market in China was also prepared.

Through the development of the country profiles, a long list of interesting REHC programmes was compiled. Twelve of these programmes were then selected based on criteria designed to identify programmes most likely to provide 'best practices'. These programmes were then investigated in-depth using publicly available information and interviews with key programme personnel. When unsuccessful or less successful programmes were identified during the course of the research, these were also noted.

Best practices were extracted from the 12 programme case studies. These best practices were analysed, and then organized and profiled in the Best Practices Guide.

Country review

Among the group of countries studied, Austria was the clear leader in solar thermal per capita total installed capacity in 2006. Denmark, Germany and Japan were ahead of the remaining countries in 2006, based on per capita total installed capacity, but were significantly below Austria. It is interesting that Austria, Denmark and Germany have relatively low annual solar insolation levels; the high number of solar collectors in these countries cannot be explained by abundant solar resources. It is also interesting to note that both Austria and Germany have high relative technology costs. It appears that policies and programmes have been instrumental in encouraging solar use in some of these countries.

Our modelling of levelized unit energy costs in the US, Denmark and Japan showed that Japan had the most competitive costs for solar technology, followed by Denmark and the US. However, even in Japan, the cost of solar heating exceeded the cost of electricity and natural gas. This suggests that solar heating continues to rely on financial incentives to overcome financial barriers in all countries.

While consistent information on solar thermal installation capacity is relatively accessible, energy-efficient biomass and heat pump installation information is difficult to find and compare. Residential energy-efficient biomass boilers, furnaces, stoves and fireplaces are typically not included in statistics. Using wood pellet production as a proxy for use of energy-efficient pellet burners is complicated by the large quantities of pellets that are exported. Likewise, both air and ground-source heat pumps do not appear to be systematically tracked in many countries. Air-source heat pumps, in particular, are not considered a renewable energy technology in some countries. Because of the lack of information, it was impossible to determine which countries are leaders in small-scale energy-efficient biomass technology. With respect to heat pumps, the available data indicate that Austria and Norway appear to be per capita ground-source heat leaders.

Programme review

During the study of successful programmes, it became clear that programme sponsors, designers and managers define 'success' in several different ways. While the goal of many programmes is to increase the penetration of one or more renewable energy technologies, other programmes focus on improving the quality of installations or increasing the use of REHC technologies in niche markets. Good programme design will depend on the definitions of success; therefore careful consideration of programme goals is important prior to the start of programme design.

Several trends were observed during the study of successful programmes:

- Long-term, continuous programmes appear to promote the most successful market transformation. This can be seen in the success that both Austria and Germany have achieved in per capita solar installations, in spite of their modest solar resources and cool climates.
- Many of the successful programmes studied were multifaceted, using several types of programme instruments to address different categories of market participants.
- Several countries that have had considerable success with incentive and education programmes have now moved to requiring the use of renewable thermal energy in new buildings and, in some cases, buildings undergoing major renovations.
- REHC programmes have variety of objectives, including increasing technology uptake, improving quality and reaching niche markets.

Best practices

Thirty-two best practices were extracted from the programmes studied. A practice or lesson learned was considered a 'best practice' if:

- It was found to be widely used in successful programmes.
- It was found to be used, often more than once, in countries that have been particularly successful at promoting renewable technologies.
- It was found in a programme that has been particularly successful and appears to have been essential to that programme's success.

Analysis of these best practices revealed that:

- The type of technology addressed by the programme is less important in determining best practices than the programme phase, market maturity or barrier faced. This suggests that the large body of experience developed in promoting residential energy efficiency may largely transfer to the promotion of residential renewable energy use.
- The best practices addressed all five categories of market barriers (acceptability, accessibility, affordability, availability and awareness). However they did not address all of the potential barriers in each category. This suggests that beyond the adoption of best practices used in other

jurisdictions, the promotion of REHC would benefit from a top-down barriers analysis and a consideration of innovative new approaches to overcome barriers.

- Some best practices are both globally applicable and essential. This group of best practices should be applied to all programmes, regardless of technology, programme type or market phase. Another group of best practices apply to most programmes and again appear to be essential to success.

- Robust programme evaluation appears to be rare, particularly when compared to typical practice in energy efficiency programmes. Evaluation needs should be considered for all residential renewable thermal energy programmes, and best practices and methodologies developed for energy efficiency programmes should be examined for transfer to renewable programmes.

The 'guide'

The Best Practices Guide in Part 2 provides information and advice on proven best practices to support the deployment of renewable energy technologies for heating and cooling in the residential sector. As noted earlier, the advice is aimed at policymakers or others in governments, agencies and utilities involved in four programme phases: portfolio planning, programme design, programme implementation and programme evaluation. Because the information is drawn from real examples (bottom-up approach), the guide does not present a systematic menu of options for overcoming all potential barriers. Rather, it is hoped that the lessons learned from real programmes can complement other sources of guidance.

The guide contains four flowcharts, one for each programme phase. To use the document, readers are instructed to turn to the flowchart that matches the relevant phase. Each flowchart is organized by market stage and programme barrier. Readers identify the market stage and programme barrier of interest, and then turn to the suggested Best Practices to find the solutions most likely to be applicable to their situation.

The 32 best practices discussed in this guide are listed in Exhibit 1. Each best practice includes a list of applicable programme phase(s), market deployment stage(s) and barrier(s). It also includes a description, additional guidance and one or more examples.

- Establish plans and programmes to achieve the established targets. These programmes should address all market participants and should be long-term
- Consider having a third party design, implement and/or evaluate the programme
- Break longer-term targets into shorter-term milestone targets
- Design and implement evaluations and refine both portfolio targets and programmes based on the evaluations
- Develop flexible plans and tools to support regional or municipal progress to the established targets
- Establish clear, consistent, ambitious but achievable portfolio targets
- Identify and encourage 'early adopters'
- Consider offering additional incentive amounts for higher quality equipment or additional efficiency measures
- Consider requesting technical drawings from installers
- Regulate the quality of the equipment and installations funded with incentives
- Design and implement long-term programmes
- Offer incentives by energy (kWh) or capacity (kW or m^2) rather than as a percentage of cost
- Penalize non-compliance with 'stick' programmes
- Develop and run programme in close cooperation with relevant public bodies and industry partners or organizations
- Establish market needs/barriers prior to programme design
- Establish clear programme objectives, targets and performance indicators
- Consider requesting an energy audit certificate with the application
- Design and implement multifaceted programmes
- Design application process to be efficient for both homeowners and programme administration
- Ensure that the market infrastructure can successfully deliver and install products being promoted
- Design programme with adequate flexibility to adapt to market changes
- Engage media to provide publicity
- Develop and distribute tools that facilitate household participation
- Invest in high-quality promotion materials for information campaigns
- Ensure a central point for accurate information
- Design programme with entry and exit strategies
- Get committed renewable energy champions to participate
- Include an evaluation component in the programme design
- Design programme to complement, not conflict, with other programmes in region
- Limit information campaigns to specific, clear messages
- Keep business interests in the background during information campaigns
- Monitor and publicize progress towards targets

Exhibit 1 Best practices

Conclusions

The programmes studied during this project are rich in best practices and lessons learned. Many of these best practices are detailed in Part 2. In general, the identified best practices are not technology-specific but rather are linked to programme phase, market stage and/or barrier faced. This suggests that the large body of experience developed in the delivery of energy conservation or photo-voltaic solar system support programmes may also inform REHC technology programme design. The one target group of technologies that requires additional consideration is energy-efficient biomass stoves or boilers, due of the need for a well-developed and reliable fuel supply chain.

Some of the identified best practices are globally applicable to all programme types and are essential to success. These include: setting clear, consistent, ambitious but achievable targets; establishing plans and programmes to achieve those targets; monitoring and reporting progress towards the targets; and designing and implementing evaluations to assess success, then refining targets and programmes. A number of additional best practices are applicable to most programme types and also appear to be essential to success.

Our study of REHC programmes also suggested that many of these programmes have not been sufficiently evaluated. This makes it difficult to verify conclusions about their success, and also may mean that programmes are not as effective and efficient as they could be. Evaluation of energy efficiency support programmes has been more robust, particularly in the context of utility demand-side management (DSM) programmes in North America. Evaluation methods and tools developed for energy efficiency programmes should be considered for transfer to REHC support programmes.

The key conclusions of this study are:

- **Appropriate best practices depend on the context.** This includes the objectives (for example, increasing deployment, improving quality, filling niches, cost-effectiveness, and so on), the authority involved and the programme development phase.
- **Appropriate best practices depend on the level of market maturity.** This means that it is important for policymakers to appreciate the level of penetration of the various technologies.
- **Best practices need to respond to the relevant barriers.** This means that policymakers need to understand the barriers faced by REHC, including both market barriers of acceptance, accessibility, affordability, availability and awareness, but also practical barriers such as lack of information, lack of capacity and lack of consistent political direction.
- **There is a wealth of experience with best practices available to be shared with policymakers.** These address many of the potential barriers and cover all barrier types. They have been captured and documented in Part 2 of this book.

- **Some barriers have not been sufficiently addressed.** Although documented best practices address many barriers, they do not address all of them. This means that there is a need for further research to understand barriers and a need for ongoing policy innovation. Fortunately, we found evidence that this innovation is happening and identified several new programmes that were too recent to be included in the scope of the present study. We also identified other areas of energy policy (for example, energy efficiency, PV support) that face similar barriers and that could be investigated for more best practices.
- **The effectiveness of best practices is unknown.** The best practices included in Part 2 were mostly chosen on the basis of subjective evidence of success. The reality is that most REHC programmes are not being evaluated with any reasonable degree of rigour. This means that the outcomes of the programmes (and by extension the associated best practices) are not being adequately measured, verified and reported.
- **Awareness of REHC best practices is uneven.** Good sources of best practice information and mechanisms for sharing information about best practices and lessons learned in other jurisdictions have not been developed.

Recommendations

Based on these conclusions, we offer the following recommendations:

- **Policymakers should begin by developing a consensus understanding of the context for the intervention.** This means understanding and clearly articulating the problem and the REHC objectives, as well as appreciating the limits of the relevant authority and the maturity of the policy development.
- **Before proposing REHC programmes, policymakers should invest in market research.** They need to understand the relative penetration of the various technologies (i.e. market maturity), as well as the barriers to market penetration (for example, in terms of the five 'A's).
- **Policymakers should make use of the *Renewable Heating and Cooling Best Practices Guide* found in Part 2.** By consulting the guide, they can find information on relevant approaches that have been used to overcome similar barriers in comparable situations.
- **Policymakers should try new approaches.** Beyond the examples in the *Best Practices Guide*, there will be additional barriers and circumstances that require different practices. Some may be borrowed from other areas of energy policy or adapted from academic and public policy research. This project focused on the bottom-up collection of best practices from real programmes, as opposed to the systematic examination of barriers and

solutions. Another study commissioned by IEA-RETD *(RENBAR – Good Practices for Solving Environmental, Administrative and Socio-economic Barriers in the Deployment of Renewable Energy Systems)* is addressing barriers more systematically but is not focused on REHC technologies. It is recommended that IEA-RETD explicitly address the links between the two studies and fill the gaps in the *Renewable Heating and Cooling Best Practices Guide*, by incorporating applicable guidance for barriers that were not addressed in this REHC project. IEA RETD should also consider follow-up research to extract best practices from the newer, more innovative programmes identified during this study.

- **Policymakers should conduct robust evaluations of their programmes.** Governments and stakeholders need credible information on the relevance, success and cost-effectiveness of REHC programmes in order to determine whether they represent appropriate investments of public resources and to determine which best practices truly contribute to success. This means adopting modern evaluation methods regarding measurement and verification, as well as dealing with attribution issues such as free ridership and spillover.

- **IEA-RETD should promote greater awareness of REHC best practices.** This includes promoting the *Renewable Heating and Cooling Best Practices Guide*, as well as monitoring its use and updating its content. It should also include promoting networks of REHC policymakers and sponsoring events that would enable such policymakers to share best practices and lessons learned. IEA-RETD should identify other regional or international organizations with similar objectives to share the burden and leverage additional intellectual capacity.

Additional recommendations for future work

Lack of comparable, systematically-collected information on residential energy-efficient biomass and heat pump installations makes it difficult to determine which regions are most successfully promoting these technologies. A system for collecting and comparing this data should be developed.

This project focused on small to medium-scale renewable thermal installations. Large-scale renewable thermal installations were not studied because they present different challenges and opportunities. A project focusing on best practices for these installations would provide guidance to programme developers, implementers and evaluators of larger-scale installations.

The Intelligent Energy Europe Res-H project is compiling more in-depth country profiles for seven countries/regions: Upper Austria; Austria, including Styria; Greece; Lithuania; Netherlands; Poland and the UK. Performing similar in-depth country profiles for the remainder of the IEA-RETD countries was beyond the scope of this project but could be considered for future work.

Notes

1   IEA-RETD, *Renewables for Heating and Cooling*, Paris: OECD/IEA, 2007. Available at www.iea.org/publications/free_new_Desc.asp?PUBS_ID=1975.
2   See www.iea-retd.org for additional information about the IEA-RETD.
3   IEA-RETD, *Terms of Reference: Innovative Policies and Markets for the Deployment of Renewable Energies for Heating and Cooling in the Residential Sector (IREHC)*, IEA, 15 October 2008.

PART 1

Best Practices in the Deployment of Renewable Energy for Heating and Cooling in the Residential Sector

1
Introduction

1.1 Background

In 2007, a report entitled *Renewables for Heating and Cooling – Untapped Potential* was jointly prepared by the International Energy Agency (IEA) and the Implementing Agreement on Renewable Energy Technology Deployment (IEA-RETD).[1] This report provided an overview of renewable energy for heating and cooling (REHC) technologies and discussed the applications, market status and research needs and priorities for these technologies. The report also included a study of current programmes supporting REHC deployment in 12 Organisation for Economic Co-operation and Development (OECD) countries and a short discussion of good practices and lessons learned for each tech type of policy instrument.

The *Renewables for Heating and Cooling – Untapped Potential* report concluded that REHC technologies can offer net savings (based on life cycle cost) compared to conventional heating systems, but that the economic viability of these systems varies greatly by location. The study found that well-designed programmes for market development do exist and that there is potential for many countries to increase their current use of REHC technologies. The study concluded with several recommendations urging governments to develop programmes to increase REHC technology use. It also recommended 'a review of best practices and lessons learned' by leading countries in developing programmes to support REHC technologies and a sector-specific analysis to evaluate which approaches are most appropriate for each of the residential, industrial, commercial and institutional sectors.[2]

Building on, and following, the recommendations of the earlier work, this current project was commissioned under the IEA-RETD to examine the REHC technology status in a set of countries, identify and investigate significant residential-sector REHC support programmes in these countries and develop and analyse a set of best practices from these programmes.

This book summarizes the work performed during the above-described project. As part of the project, a stand-alone companion document, *Renewable Heating and Cooling Best Practices Guide*, was developed for programme developers at all levels. This latter document makes up Part 2 of this book.

1.2 Objectives

The overall objective of this project was to develop materials 'to assist policymakers in developing robust, effective, sustainable programmes that can overcome the most common barriers that occur in deploying REHC in the residential sector'.[3] More specific objectives were:

- To assess REHC programmes in the residential sector and identify those that have been the most effective.
- To profile the resulting best practices in portfolio planning and programme design, implementation and evaluation.
- To create a guide of practical steps and methods to implement the best practices.
- To effectively communicate and disseminate the best practices and guidance to appropriate audiences, including policymakers.

1.3 Project scope

As illustrated in Exhibit 2, the scope of this project is defined in several dimensions: audience level of authority, programme phase, stage of market maturity, barrier to deployment, instrument category, target technology and applicable sector. The dimensions inside the dashed box: programme phase, market maturity stage and barriers to deployment, were used to organize the best practices in Part 2.

1.3.1 Audience level of authority

The Best Practices Guide (Part 2) developed during this project is intended for personnel responsible for renewable energy portfolio planning and programme development, implementation and evaluation at all levels of authority: multi-national, national, regional and local.

1.3.2 Programme phases

REHC programme design and implementation can be divided into four phases:

1 *Portfolio planning* encompasses initial problem definition and goal clarification, setting objectives, consideration of options and alternatives, identifying intervention points, identifying stakeholders and conducting consultations (if necessary), and selecting instruments.
2 *Programme design* can be done by the funding agency or a third party. Ideally, the programme design is based on the work done during the portfolio planning phase.
3 *Programme implementation* can also be done by the funding agency or a third party. During this phase, the programme is implemented.

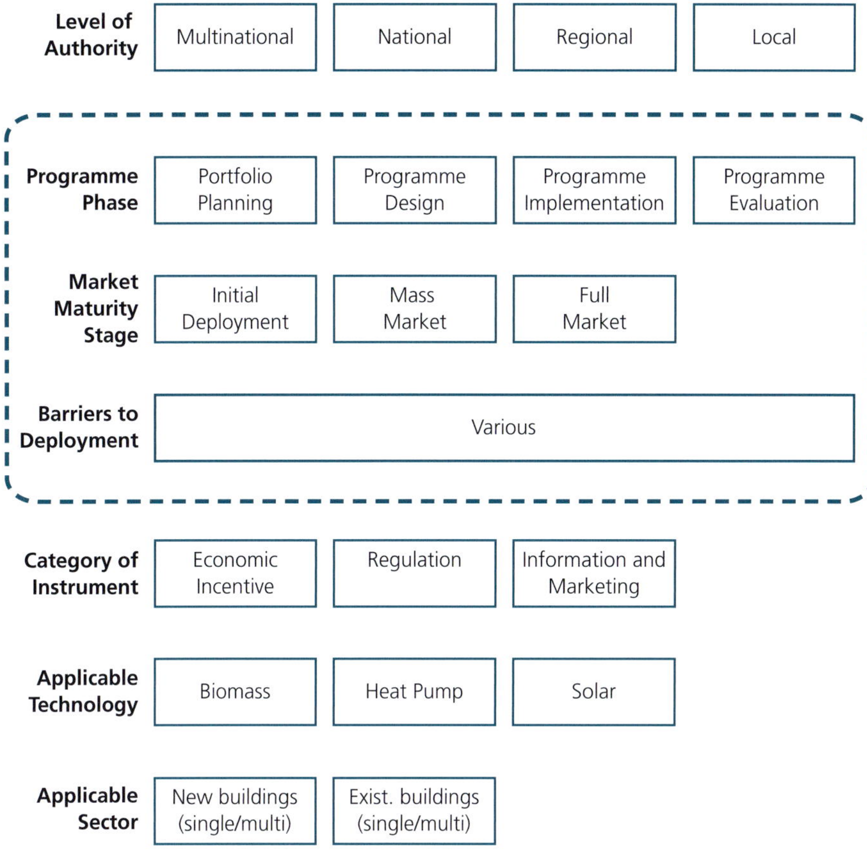

Exhibit 2 Project scope dimensions

4 *Programme evaluation* closes the loop by assessing progress against objectives, and reassessing both programme logic and processes. The programme is then modified as required. Evaluation is usually best done by individuals not involved in the programme design and implementation.

This project targeted all four programme phases.

1.3.3 Market maturity stages

Programmes designed to develop REHC markets can be categorized in the following manner:

- *'Initial deployment' programmes* are appropriate for a market that is the initial stage. In this type of market, the public is largely unaware of the technology, the technology is not readily available and installation and maintenance support is difficult to find. REHC programmes for this market stage include demonstration and pilot projects. Training for retailers and installers is also appropriate.

- *'Mass market' programmes* are aimed at a market that has passed through the initial deployment stage. The technology is becoming better known to the public, is being sold by retailers, and competent installers can be readily found. Appropriate REHC programmes include 'carrot' programmes, such as economic incentives, and 'guidance' programmes to educate both the public and industry members. In some specific situations, such as under certain political conditions, 'stick' measures may be appropriate, although these are more typically found under 'full market' conditions.
- *'Full market' programmes* target well-developed markets. In these markets, the public is knowledgeable about the technology and its benefits, the technology can be easily obtained and an adequate number of well-trained installers are in place. 'Carrot' measures may still be appropriate, but 'stick' measures such as mandatory installation regulations may be appropriate for bringing non-participants into the market.

Note that it is not unusual for different REHC technologies to be in different market stages in a given geographical area.

This project targeted all three market maturity stages; however, most programmes examined were aimed at either 'mass market' or 'full market' stages.

1.3.4 Barriers to deployment

REHC technologies face a variety of barriers to wide-spread use. These barriers can be categorized in a number of ways. One method uses five 'A's:

1 *Acceptability:* Some REHC technologies may not (yet) be as easy to use as conventional technologies. Maintenance costs might be higher, reliability might not be as good or use may not be as convenient.
2 *Accessibility:* Laws, regulations or codes may make REHC installation or use difficult. For example, building codes may require expensive inspections of solar water collectors on roofs.
3 *Affordability:* Many REHC technologies are more expensive to install than conventional technologies. Financing for REHC technologies may be harder to acquire than for conventional technologies.
4 *Availability:* REHC technologies may have limited availability or lack qualified installers.
5 *Awareness:* The benefits of REHC technologies may not be well known, or potential users may not know how to obtain the technology.

Programme best practices vary not only by programme development phase and market stage, as described above, but also by barrier.

1.3.5 Instrument categories

Methods of involving public and industry in REHC programmes can be divided into three categories: economic incentives, regulations and information and marketing. Another common method of categorizing support instruments is with the descriptive terms 'carrots', 'sticks' and 'guidance'. These categorization methods often, but not always, correspond to each other in a one-to-one manner.

- *Economic incentives* (often called 'carrots') include grants, loans, rebates and tax credits. Economic disincentives such as fines, on the other hand, would be 'sticks' (see below).
- *Regulations* (often called 'sticks') include building codes and laws requiring minimum installations. In some cases, regulations can be 'carrots' in that they make installing a REHC technology easier.
- *Information and marketing* (often called 'guidance') includes public awareness and education campaigns, quality assurance standards, support resources and training.

All three instrument categories were examined in this project.

1.3.6 Target technologies

The target technologies for this project are:

- Active solar thermal for air and water heating or cooling.
- Biomass (pellets, wood and wood waste).
- Geothermal (ground-source heat pumps).
- Heat pump technologies based on ambient air heat (air-to-air and air-to-liquid).

Active solar thermal systems

Active solar thermal systems can use either glazed or unglazed solar collectors.[4] Unglazed collectors are employed in applications such as swimming pools and crop drying. Because this report is focused on the use of renewable thermal energy for space heating and domestic hot water, unglazed collectors have not been examined.

Glazed solar collectors can be divided into two major types: flat-plate collectors and evacuated tube collectors. Flat-plate collectors are appropriate for lower demand hot water systems, where the higher water temperatures produced by evacuated tube collectors could be dangerous. They are also often more appropriate when snow accumulation is a problem, because the heat loss through a flat plate collector will often melt the snow. Evacuated tube collectors can produce higher temperatures in the working fluid and therefore are often more appropriate for constant high demand water heating systems or process loads. Flat-plate collectors have historically been less expensive than evacuated tube collectors, however, less expensive Chinese evacuated tube collectors can now be purchased. Concern has been expressed about the quality of the Chinese products but several models have been approved by the Solar Keymark programme. None of the policies or programmes analysed in this report distinguishes between these two technologies. Costs were analysed assuming a flat-plate collector.

Small biomass systems

For this project, small biomass systems are assumed to burn wood pellets, wood or wood waste. Although biomass systems may be supplied with other fuel types, these are less common at the residential scale. Systems may be boilers, stoves or furnaces. Support programmes typically have minimum efficiency, and possibly emissions, requirements.

Heat pumps

Heat pumps can either be ground-source or air-source:

Ground-source (or geothermal) heat pumps are the most common method of using geothermal heating and cooling. When heating is required, ground-source heat pumps move heat from the ground or ground water to the conditioned space. When cooling is required, the pumps are reversed. The ground loop or connection can take several forms, depending on the local geology and environmental concerns. Because the ground temperature is fairly constant throughout the year, these systems are typically more efficient that air-source heat pumps, particularly in cold climates, but are also more expensive to install.

Air-source heat pumps move heat from the ambient air to the conditioned space when heating is required, and are reversed when cooling is required. These heat pumps may be classified into air-to-air and air-to-liquid heat pumps. Air-to-air heat pumps heat or cool air and are suitable for new construction or in retrofit projects in homes already using air ducts for heating or cooling. They are also well-suited to providing air-conditioning. Air-to-liquid heat pumps heat or chill water and are suitable for houses that use a hydronic heating system (typically with radiators or radiant floor heating.) Condensation can be a problem when cooling with air-to-liquid heat pumps. On the other hand, the heating efficiency for air-to-liquid heat pumps may be higher than that for air-to-air heat pumps when in-floor heating can be done. The choice between these two heat pumps types depends on the existing or planned heat distribution system and the need for cooling. Certain programmes studied during this project distinguish between the two types of heat pumps in order to target different types of homes.

1.3.7 Target Sectors

Solar thermal technologies, heat pumps and biomass technologies are often small to medium-scale technologies heating or cooling a single dwelling, or, at most, a few dwellings. However, some of these technologies can be used in large-scale applications to provide district heating. The two types of applications present different challenges, as shown in Exhibit 3 below:

Because the differences outlined in Exhibit 3, best practices for programmes targeting small- to medium-scale applications will not be the same as best

Challenge	Small- to Medium-Scale	District Heating
Body to motivate	Homeowner, or possibly groups of homeowners, owners of multi-unit dwellings, or construction firms working with residential building owners.	Utility (private or government) or private corporation.
Motivation	Likely to be environmental concerns. May not apply strict financial criteria.	If private, more likely to require investment meet short-term business investment criteria. If government, able to focus on long-term benefits.
Financing	Personal savings or personal loan.	Business or government budget or loan.
Technical	Small-scale generation and very local distribution.	Large-scale, with significant distribution system.
Marketing	Must address and motivate many individuals with a variety of situations and motivations.	Can target a few organizations with customized messages.
Operation and maintenance	Operation and maintenance must not require specialized skills or expertise must be readily available in community.	Owner may have technical staff that can be trained in operation and maintenance.
Market infrastructure	Technology and installation must be readily available and visible in community.	Technology and installation will be specialized and can come from a distance.

Exhibit 3 Small- to medium-scale vs. district REHC applications

practices for programmes targeting district heating or cooling systems. This project focused on developing best practices for programmes aimed at influencing homeowner behaviour. Although 'homeowner' may imply an individual or family living in and owning a single, detached home, many of the programmes reviewed in this report also targeted groups of homeowners, owners of multi-unit dwellings and construction firms working with building owners. We recommend developing best practices for programmes aimed at district heating owners in a future project.

In addition to the issue of scale, type of dwelling may also be an issue. Programmes may address single, detached houses in a different manner than they address multi-unit dwellings. Alternatively, programmes may differentiate between new construction and existing buildings. Where these distinctions are important, they will be discussed during the programme review.

1.4 Approach

This project was divided into four steps, as shown in Exhibit 4.

The approach for each step is described below.

1.4.1 Step 1: Country review

During Step 1 of this project, we developed country summaries for 14 countries: the ten IEA-RETD member countries (Canada, Denmark, France, Germany, Ireland, Italy, Japan, Netherlands, Norway and the UK); Austria and Spain, due

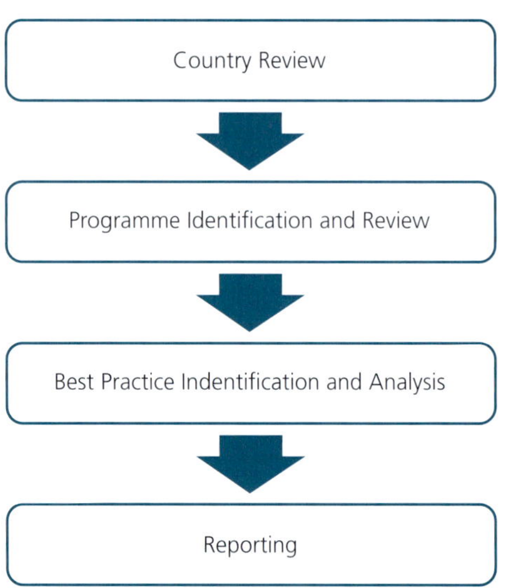

Exhibit 4 Project steps

to their success with renewable energy; the US, for both programme variety and success; and China, due to its rapidly developing solar thermal market. These summaries provided background information for further investigation and evaluation of promising programmes. They also allowed a comparison between countries to identify leaders in REHC implementation. Finally, we compared, to the extent possible with the available data, the relative progress each country has made in using REHC technologies.

1.4.2 Step 2: Programme identification and review

Following the country research, we examined 12 successful programmes in-depth. We used the following criteria to choose the programmes to be examined:[5]

- Evidence that programme likely caused a significant increase in renewable technology uptake based on either:
  - Programme evaluation reports.
  - Country or region has either high installed capacity or a recent high rate of installed capacity under circumstances that are not all favourable. Under these conditions, it is likely that programmes are responsible for the high uptake.
- Programme has been unusually stable and continuous over a period of time.
- Programme has unique features not found in other programmes being studied.
- Enough historical data is available to determine how well the programme is working and what lessons have been learned.
- Programme is of sufficient scale (time, region covered or budget) to have made an impact.
- Programme selected from the most successful countries, as determined in during the country research phase of the study.
- Programme provides diversity among programmes chosen, in terms of
  - Technology(ies) targeted.
  - Type of programme.
  - Dwelling types and markets targeted.

Once the target programmes were chosen, in-depth information was collected from a combination of publicly-available information sources and interviews with programme personnel.

Based on the assembled programme information, we identified patterns of success by observing common traits of successful programmes. We then examined how various programmes were defining success, and determined what drivers and indicators corresponded to those definitions.

1.4.3 Step 3: Best practice identification and analysis

During Step 3, we developed and then analysed a list of 'best practices', based on the information developed in Steps 1 and 2. This work was carried out in the following manner:

- Best practices and lessons learned were collected from the detailed programme descriptions built during Step 2. These came from programme evaluations, where available; programme administrators; and our own analysis of the programme.
- For each best practice, the following were elaborated: applicable programme phase, applicable market deployment stage, barriers addressed, description, additional guidance, and examples. Where additional positive or negative examples were available, or where outside reports could be referenced, this material was included.
- Because many of the programmes examined did not have full evaluations, additional information was provided on programme evaluation best practices, based on our previous experience.

1.4.4 Step 4: Reporting

During Step 4, the content of this book was created through the following sub-tasks:

- To ensure a consistent and logical presentation, an organizational framework was first developed.
- Best practices developed in Step 3 were organized according to the framework and reported in the *Renewable Heating and Cooling Best Practices Guide*.
- This report was compiled based, in part, on interim reports developed earlier in the project.

1.5 Report organization

The remainder of Part 1 is organized as follows:

- Chapter 2 presents a comparison of the experience of 14 countries with REHC technologies. The detailed country summaries on which these comparisons are built can be found in Chapter 6.
- Chapter 3 explains how REHC programmes were selected for further investigation; summarizes programme trends; and discusses definitions, drivers and indicators of success. The detailed programme profiles can be found in Chapter 7.

- Chapter 4 discusses the selection and organization framework for the best practices guide. It also provides a general best practice analysis. The full *Renewable Heating and Cooling Best Practices Guide* is included as Part 2 of this book.
- Chapter 5 provides conclusions and recommendations for further research.

Notes

1   See www.iea-retd.org for additional information about the IEA-RETD.
2   IEA-RETD, *Renewables for Heating and Cooling* (Paris: OECD/IEA, 2007). Available at www.iea.org/publications/free_new_Desc.asp?PUBS_ID=1975.
3   IEA-RETD, *Terms of Reference: Innovative Policies and Markets for the Deployment of Renewable Energies for Heating and Cooling in the Residential Sector (IREHC)* (IEA, 2008).
4   Although passive solar thermal systems can provide residential heating, these systems are not included in IEA statistical data and were therefore also not included in the earlier report *(Renewables for Heating and Cooling – Untapped Potential)*. This project has also omitted passive systems due to lack of data.
5   Note that these criteria do not define 'successful' or 'unsuccessful' programmes and policies; they provided guidance on which programmes and policies were likely to be worthy of additional study.

2

Review of Country Experience

As shown in Exhibit 4, the first step in this project was to review country experience with REHC technologies. Information for each country was collected, as described in Section 2.1. A number of comparisons were then made, such as:

- Gross domestic product and conventional energy costs and availability.
- Progress on solar, biomass and heat pump installations.
- Cost of renewable and conventional energy costs.

These comparisons are presented in Sections 2.2–2.5.

2.1 Information collected

Context, market status, programmes and results achieved were collected for each of the 14 countries reviewed.[1] The countries include the ten IEA-RETD member states: Canada, Denmark, France, Germany, Ireland, Italy, Japan, the Netherlands, Norway and the UK. In addition, Austria and Spain were included due to their success with renewables, and the US was collected because of both the variety and success of their renewable energy activities. Finally, China was also briefly reviewed because the large developing solar hot water heating market developing in that company.

2.2 Comparison of GDP and residential conventional energy costs

Exhibit 5, below, presents the gross domestic product (GDP) adjusted by purchasing power parity (PPP)[2] per capita, adjusted electricity and natural gas prices, and qualitative indications of the relative fuel cost and domestic availability of conventional energy. The relative availability of domestic conventional fuel was determined qualitatively from the information presented in the country summaries.

Country	GDP (PPP) Per Capita (2008 est)[3]	Electricity Cost[4] (€/kWh)	Electricity Cost per million € of GDP (PPP)	Natural Gas Cost[5] (€/GJ)	Natural Gas Cost per thousand € of GDP (PPP)	Relative Fuel Cost Price Comparison (elec/natural gas)	Domestic Availability of Conventional Energy[6]
Austria	€30,462	0.18	5.84	22.99	0.755	M/H	L
Canada	€30,923	0.08[7]	2.59	7.76[8]	0.251	L/L	H
Denmark	€29,923	0.26	8.81	36.89[9]	1.23	H/H	H
France	€25,154	0.12	4.82	14.46	0.482	M/M	M
Germany	€26,769	0.10	3.91	17.81	0.665	M/H	M
Ireland	€36,769	0.18	4.81	15.09	0.410	M/M	L
Italy	€23,846	0.21[10]	8.72	17.47	0.733	H/H	L
Japan	€27,154	0.15[11]	5.52	13.11[12]	0.483	M/M	L
Netherlands	€31,780	0.17	5.44	19.37	0.610	M/H	M
Norway	€44,231	0.16	3.71	NA	NA	M/NA	H
Spain	€26,231	0.14	5.21	15.98	0.609	M/H	L
UK	€28,692	0.15	5.08	10.99	0.383	M/M	M
US	€36,923	0.07[13]	1.90	7.82[14]	0.212	L/L	H

Exhibit 5　GDP, residential conventional energy prices and domestic availability of conventional energy

The countries with the highest relative fuel costs are Austria, Denmark, Germany, Italy, the Netherlands and Spain. The availability of domestic conventional fuel sources does not correlate well with relative energy prices, perhaps because countries vary widely in their energy tax policies and the presented fuel costs are after-tax values.

2.3 REHC installed capacity, resource and cost comparisons

In subsections 2.3.1–2.3.3, solar, biomass and heat pump installations in the various countries are compared, when possible, along with external factors such as resource availability and price advantage. These comparisons provide, at a glance, an indication of the success achieved in each country, together with an assessment of the influence of both external and programme factors. Countries that are making progress in spite of few external (natural or non-policy) advantages are good candidates for having adopted best practices.

2.3.1 Solar

Solar installations and programmes have been well-studied and data on total installed capacity, yearly installed capacity and costs are all readily available in forms that allow for easy comparisons between countries. This information has been collected and compared in Exhibit 6. Exhibit 7 graphically compares the total per capita installation.

Among the group of countries studied, Austria was the clear leader in per capita total installed capacity in 2006. Denmark, Germany and Japan were ahead of the remaining countries in 2006 based on per capita total installed capacity, but were significantly below Austria. The capacities installed in 2007 in Denmark and Japan are small in comparison to the capacity installed previously, suggesting that, while these countries have been solar leaders in the past, they are no longer in that category.

It is interesting that Austria, Denmark, and Germany have relatively low annual solar insolation levels; the high number of solar collectors in these countries cannot be explained by abundant solar resources. It is also interesting to note that both Austria and Germany have high relative technology costs.

2.3.2 Biomass

Exhibit 8 presents and compares information about residential energy-efficient biomass use in the selected countries. Information on the number of installed residential energy-efficient biomass devices is not readily available, as was also noted in the *IEA report Renewables for Heating and Cooling: Untapped Potential*. Some information on the installed number of pellet burners is available and is included in Exhibit 8, however, it is not possible to determine which countries

Country	Success Indicators[15]			External Factors			
				Access to resource			
	Total installed capacity at the end of 2006 (MWth)	Total installed capacity per person at the end of 2006 (kWth/1000 inhabitants)	Capacity installed in 2007 (MWth)	Annual global horizontal irradiation (kWh/m^2)[18] (multiple values where available	Rating[19]	Relative conventional energy cost (elec/natural gas)[16]	Relative cost of technology[17]
Austria	1898.1	230.5	196.7	1108	L	M/H	H
Canada	57.9	1.8	2.7	1273, 1377	M	L/L	M
Denmark	261.7	48.2	16.4	985	L	H/H	M
France[20]	745.9	12.0	222.4	1057, 1540	M	M/M	M
Germany	5637.8	68.2	658.0	999, 1143	L	M/H	H
Ireland	11.1	2.7	13.8	948	L	M/M	M
Italy	512.0	8.8	171.5	1251, 1552	H	H/H	M
Japan	4747.0	37.1	119.2	1168, 1350	M	M/M	M
Netherlands	221.8	13.6	13.9	1045	L	M/H	M
Norway	7.4	1.6	0.5	967	L	M/NA	L
Spain	663.4	15.4	183.4	1446, 1754	H	M/H	M
UK	175.6	2.94	37.8	890, 955	L	M/M	H
US	1634.1	5.48	105.9	1487, 1816	H	L/L	L

Exhibit 6 Comparison of solar heating systems (glazed or evacuated tube installations)

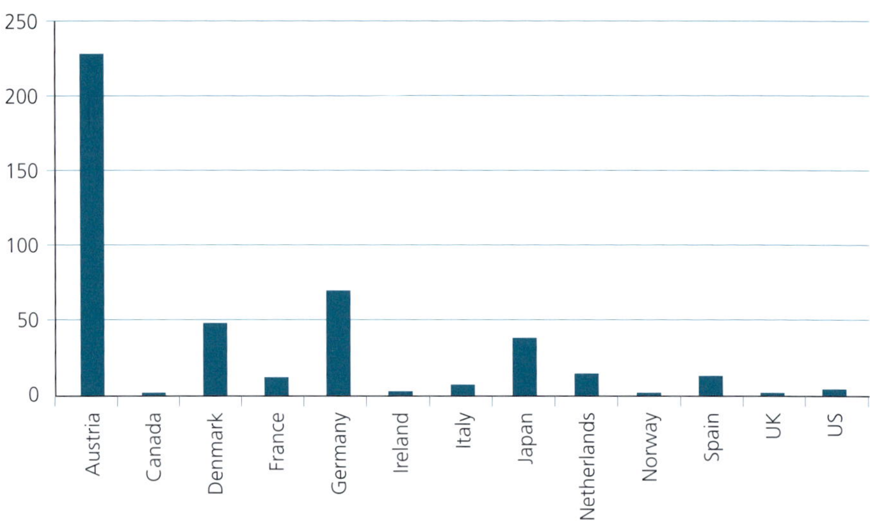

Exhibit 7 2006 total per capita installed solar thermal capacity (kWth/1000 inhabitants)

have the highest per capita number of energy-efficient biomass installations, nor which countries have the highest recent annual installation rate. Information on annual quantities of pellet production is more readily available, but can be misleading because some countries, such as Canada, export large percentages of their domestic pellet production.

Because of the lack of information, it is impossible to determine which countries are leaders in small-scale energy-efficient biomass technology.

2.3.3 Heat pumps

Information on air-source and ground-source heat pumps is presented in Exhibit 9. Comparable numbers of residential air-source heat pump installations, either total or annual, is difficult to find, consequently the data presented on air-source heat pumps installations in Exhibit 9 is drawn from a variety of sources. The lack of data may be explained by the fact that air-source heat pumps are not considered renewable technologies in all countries, and therefore the topic appears not to have been comprehensively studied.[21]

Readily comparable information on ground source heat pump installations was only slightly more available. Some of this information is presented in Exhibit 9, although, as noted in some country profiles in Chapter 6, this information is often contradictory. In the countries studies, and for which there is data, Austria and Norway appear to be per capita ground-source heat pump leaders. The per capita installations of GSHPs, where available, are presented in Exhibit 10.

	Success Indicators	External Factors				
		Access to resource				
Country	Number of Pellet Burners[22]	Biomass resources, wood[25]	Domestic pellet industry [million tonnes][26]	Biomass resources, pellets[27]	Relative conventional energy cost (elec/natural gas)[23]	Relative cost of Technology[24]
Austria	63000 [2008]	H	0.626 [2008]	H	M/H	H
Canada	unknown	H	1.4 [2007]	H	L/L	L
Denmark	30000 [2006][28]	M	0.070 [2008]	L	H/H	M
France	38,000 pellet stoves plus more than 7000 boilers installed in total [2007][29]	M	0.240 [2008]	M	M/M	M
Germany	140,000 [2008]	M	1.460 [2008]	H	M/H	H
Ireland	2500 [2008]	M	0.015 [2008]	L	M/M	M
Italy	800,000 [2008][30]	M	0.650 [2008]	H	H/H	L
Japan	negligible	L	0.100 [2007]	L	M/M	N/A
Netherlands	negligible	L	0.120 [2008]	L	M/H	N/A
Norway	negligible	H	0.035 [2008]	L	M/NA	L
Spain	unknown	L	0.163 [2008]	M	M/H	M
UK	negligible	L	0.125 [2008]	L	M/M	L
US	800,000 [2008][31]	H	1.2 [2007]	H	L/L	L

Exhibit 8 Comparison of energy efficient biomass installations

Country	Success Indicators			External Factors			
				Access to resource			
	Number of ASHPs installed/sold in 2007[32]	Number of GSHPs installed/sold in 2007[33]	Total number of GSHPs in use in[34]	Climate is suitable for air-source heat pumps[37]	Low-temperature geothermal resource	Relative conventional energy cost (elec/natural gas)[35]	Relative cost of technology[36]
Austria	2110	8288	40,549	M	Good	M/H	H
Canada	N/A	9103[38]	31,000 [2007 estimate][39]	Varies	Varies[40]	L/L	L
Denmark	N/A	N/A	11,250	M	Good	H/H	M
France	51,000	18,600	102,456	M	Good	M/M	H
Germany	18,389	26,887	115,813	M	Good	M/H	M
Ireland	N/A	2608	7578	M	Good	M/M	M
Italy	28,901	N/A	7500	H	Good	H/H	N/A
Japan	N/A	N/A	1330[41] [2006]	M	Good	M/M	L
Netherlands	N/A	3529	15,230	M		M/H	M
Norway	67,200	3318[42]	15,000[43]	L	Good	M/NA	L
Spain	N/A	N/A	2200[44] [2005]	H	Good	M/H	M
UK	750	3000	5350	M	Good	M/M	L
US	N/A	50,000 [2007 estimate][45]	N/A	Varies	Good	L/L	L

Exhibit 9 Comparison of heat pump installations

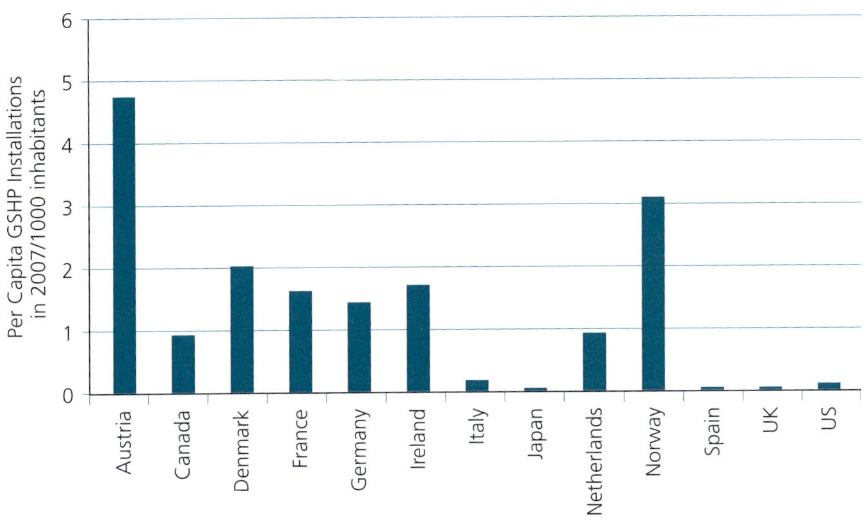

Exhibit 10 2007 per capita GSHP installations (number/1000 inhabitants)

2.4 Relative cost of renewable heating and cooling

Absent from the exhibits above are the costs of generating energy with the target renewable heating technologies. These values are very difficult to calculate due to the variations in system types, equipment and labour costs. For example, while the cost of a cubic metre of natural gas can easily be determined for a given location, the cost of installing a solar hot water system will vary depending on the type of equipment purchased and the modifications required to the existing house. To provide a rough picture of these costs, we have calculated approximate levelized unit energy costs for solar water heating systems in three countries.

The levelized unit energy cost (LUEC) represents the life cycle cost of producing one unit of energy from a given technology. Three countries were selected to illustrate the LUEC analysis of solar heating technology based on two criteria: domestic availability of the solar resource, and conventional energy. Denmark was selected because it has a relatively low solar resource and high domestic energy costs. The US, on the other hand, has a large solar resource, particularly in the south, and relatively low energy costs. Finally, Japan was chosen as a third country for analysis because it has a moderate solar resource and medium relative costs. This analysis is presented tabular form in Exhibit 11 and in graphic form in Exhibit 12.

These three examples show solar energy costs ranging from €0.17–€0.40/kWh, which in all cases was higher than for conventional energy by €0.02–€0.14/kWh. In terms of an average family home, this would result in an increase in cost over a

Country	Reference Climate[46]	Solar Heating Technology				Conventional Energy Costs[47]	
		Typical Solar Hot Water System Size (m²)[48]	Heat Delivered per year (MWh)[49]	Typical Technology Cost (€)[50]	LUEC (€/kWh)[51]	Electricity Price (€/kWh)	Natural Gas (€/kWh)
Denmark	Copenhagen	4	1.4	4350[52]	0.40	0.26	0.13
Japan	Tokyo	4	2.2	2900[53]	0.17	0.15	0.05
US	Los Angeles	6	2.6	3450	0.17	0.07	0.03

Exhibit 11 Solar and conventional energy cost in three countries

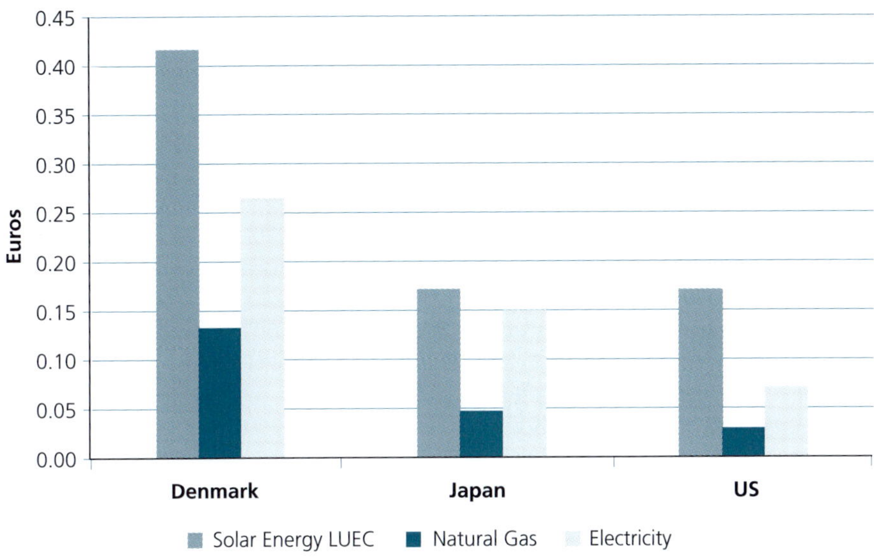

Exhibit 12 Levelized unit energy cost of solar heating in three countries (€/kWh)

conventional system of €50–350 per year, depending on whether the conventional system uses electricity or natural gas and the unitary cost of each fuel.

An increase in cost of €350 per year indicates that the investment could not be justified economically and therefore an incentive would be required to increase the solar water heating capacity of a country. An increase in cost of €50, however, may be acceptable cost to homeowners for the social and environmental benefits that they may experience as a result of purchasing a solar hot water heating system.

Notes

1   Country profiles are available in Chapter 6.
2   'Gross domestic product adjusted by purchasing power parity' takes into consideration the price of goods traded with a country (based on a specific 'basket of goods') rather than considering only the export value and exchange rate of the country's currency.
3   Values are from the *CIA World*, www.cia.gov/library/publications/the-world-factbook/rankorder/2119rank.html, accessed June 2009.
4,5 European fuel prices are for the first half of 2008 and were taken from *Eurostat*, www epp.eurostat.ec.europa.eu, accessed February 2009.
6   Qualitatively evaluated from the country summaries.
7   Marbek in-house information – Ontario average but can be much higher in northern communities
8   Marbek in-house information – Ontario average.

9   Prices for the second half of 2007, from *Eurostat*, www.epp.eurostat.ec.europa.eu, accessed February 2009.

10   Provisional value, as provided in *Eurostat*, www.epp.eurostat.ec.europa.eu, accessed February 2009.

11   *Press Release: Fuel Cost Adjustment of Electricity Rates for April 2009*, Kyushu Electric Power's Mission, www.kyuden.co.jp/en_press_h090129b-1.html, accessed February 2009.

12   *Natural Gas Price Divergence*, Platts.com news feature, 2008, www.platts.com/Natural%20Gas/Resources/News%20Features/gasdivergence/chart1.xml, accessed February 2009. May not include tax.

13   *Energy Overview*, Energy Information Administration, 2007, www.eia.doe.gov/cneaf/electricity/epm/table5_6_a.html#_ftnref1, accessed February 2009. Price varies greatly throughout the country.

14   *Average Residential Gas Consumption and Cost 1996–2007*, American Gas Association, www.aga.org/NR/rdonlyres/5778DC19-E11C-4BB4-A121-F961A613B656/0/0902Table96.pdf, accessed February 2009. Prices from 2007.

15   Weiss, W. et al, *Solar Heat Worldwide: Markets and Contribution to the Energy Supply 2007*, 2009 edition, IEA Solar Heating and Cooling Programme, www.iea-shc.org/publications/statistics/IEA-SHC_Solar_Heat_Worldwide-2009.pdf, accessed August 2009. Also Weiss, Bergmann and Faniger, *Solar Heat Worldwide: Markets and Contribution to the Energy Supply 2006*, 2008 edition, IEA Solar Heating and Cooling Programme, www.iea-shc.org/publications/statistics/IEA-SHC_Solar_Heat_Worldwide-2008.pdf, accessed January 2009.

16   From Exhibit 4.

17   A solar system costing less than 10 per cent of the GDP (PPP) per capita including installation was considered low cost (L). A solar system costing between 10–15 per cent of the GDP (PPP) per capita including installation was considered medium cost (M). A solar system costing more than 15 per cent of the GDP (PPP) per capita including installation was considered high cost (H). Technology costs were collected from catalogues, retailers and case studies.

18   Gaiddon, B. and Jedliczka, M., *Compared Assessment of Selected Environmental Indicators of Photovoltaic Electricity in OECD Cities*, IEA – Photovoltaic Power Systems Programme, 2006. Available at www.risoe.dk/rispubl/NEI/nei-dk-4726.pdf.

19   Irradiation below 1200kWh/m^2 was considered a low solar resource (L). Between 1200–1400kWh/m^2 was considered a medium solar resource (M). Above 1400kWh/m^2 was considered a high solar resource (H).

20   Includes values from some overseas departments.

21   As indicated in Exhibit 9, not all countries have a climate that is suitable for air-source heat pump use. Air-source heat pumps will not work efficiently at low ambient temperatures. Back-up heating is required, and often this back-up heating is provided by electric resistance heating. Where there is abundant hydroelectricity, this may be acceptable, but in most regions electrical heating is discouraged.

22   Pellets @las website, www.pelletsatlas.info, accessed June 2009, for Austria, Germany, Ireland, Netherlands and the UK.

23   From Exhibit 3.

24   A biomass system costing less than 25 per cent of the GDP (PPP) per capita was considered low cost (L) including installation. A biomass system costing between 25–45 per cent of the GDP (PPP) per capita was considered medium cost (M) including installation. A biomass system costing more than 45 per cent of the GDP (PPP) per capita was considered high cost (H) including installation. Technology costs were collected from catalogues, retailers and case studies.

25   Qualitative evaluation based on country summaries.

26   Most values from Intelligent Energy Europe, *Pellet market data 2008: Survey Data Collection*, Pellet @las, June 2009. Available at www.pelletsatlas.info/pelletsatlas_docs/showdoc.asp?id=090603135649&type=doc&pdf=true. For Japan, Canada and the US, from Swaan, J. and Melin, S., *Wood Pellet Export: History – Opportunities – Challenges, Small Wood 2008*, Wood Pellet Association, www.forestprod.org/smallwood08swaan.pdf, accessed June 2009.

27   Production of less than 0.15 million tonnes of pellets is considered a low resource (L). Production between 0.15–0.50 million tonnes of pellets is considered a medium resource (M). Production of more than 0.50 million tonnes of pellets is considered a high resource (H).

28   Pellet centre website, http://pelletcentre.info/resources/1010.pdf, accessed February 2009.

29   *Enquête sur les ventes d'appareils domestiques de chauffage au bois en 2007*, Paris: Observ'ER, 2008.

30   Egger, C. and Oehlinger, C., 'Burning issues: An update on the wood pellet market', *Renewable Energy World*, 7 April 2009. Available at www.renewableenergyworld.com/rea/news/print/article/2009/04/burning-issues-an-update-on-the-wood-pellet-market.

31   Niebling, C., *BioPellets: Clean, Efficient Heating with Renewable Energy from America's Forest Resources*, New England Wood Pellet, LLC, 2008, http://files.eesi.org/niebling.pdf, accessed February 2009.

32   A reliable source for total installed air source heat pump capacity per country was not available. The number of installations done in 2007 for many countries is available from: *Outlook 2008: European Heat Pump Statistics/Summary*, Brussels: European Heat Pump Association, 2008.

33,34 Unless otherwise indicated, values are from 'Heat pumps barometer', *SYSTÈMES SOLAIRES: Le journal des énergies renouvelables*, no 193, 2009. Available at www.eurobserv-er.org/pdf/baro193.pdf.

35   Taken from Exhibit 4.

36   A heat pump system costing less than 25 per cent of the GDP (PPP) per capita was considered low cost (L) including installation. A heat pump system costing between 25–45 per cent of the GDP (PPP) per capita was considered medium cost (M) including installation. A heat pump system costing more than 45 per cent of the GDP (PPP) per capita was considered high cost (H) including installation. Technology costs were collected from catalogues, retailers and case studies.

37   This was qualitatively evaluated from country summaries – a highly suitable climate does not have a cold winter (for example, Mediterranean countries).

38,39 Personal communication with Al Clark, programme manager for the Renewable Heat Programme of Natural Resources Canada.

40   Not suitable for areas with permafrost.

41   Assuming an average GSHP capacity of 10kW from 13.3MW capacity, from IEA-RETD, *Renewables for Heating and Cooling: Untapped Potential*, Paris: OECD/IEA, 2007. Available at www.iea.org/textbase/nppdf/free/2007/Renewable_Heating_Cooling.pdf.

42   Data compiled by NOVAP, the Norwegian Heat Pump Association, www.novap.no.

43   Stene, J., *Large-Scale Ground-Source Heat Pump Systems in Norway*, Presented at IEA Annex 29 Workshop, Zürich, 19 May 2008. Available at www.annex29.net/extern/09_STENE_Zuerich_Annex29_GSHPs.pdf.

44   Assuming an average GSHP capacity of 10kW from 22MW capacity from IEA-RETD, *Renewables for Heating and Cooling: Untapped Potential*, Paris: OECD/IEA, 2007. Available at www.iea.org/textbase/nppdf/free/2007/Renewable_Heating_Cooling.pdf.

45   Hughes, P. J., *Geothermal (Ground-Source) Heat Pumps: Market Status, Barriers to Adoption, and Actions to Overcome Barriers*, Oak Ridge: Oak Ridge National Laboratory, 2008. Available at www.eere.energy.gov/geothermal/pdfs/ornl_ghp_study.pdf.

46   As recommended in Weiss, Bergmann and Faniger, *Solar Heat Worldwide: Markets and Contribution to the Energy Supply 2006*, 2008 edition, IEA Solar Heating and Cooling Programme. Available at www.iea-shc.org/publications/statistics/IEA-SHC_Solar_Heat_Worldwide-2008.pdf.

47   Taken from Exhibit 4. Natural gas prices are converted to per kilowatt equivalents for ease of comparison.

48   Weiss, Bergmann and Faniger, *Solar Heat Worldwide: Markets and Contribution to the Energy Supply 2006*, 2008 edition, IEA Solar Heating and Cooling Programme. Available at www.iea-shc.org/publications/statistics/IEA-SHC_Solar_Heat_Worldwide-2008.pdf.

49   Calculated using RETScreen assuming glazed flat-plate collector of the indicated size at the reference location. Climate readings from airports in the reference location were used.

50   Marbek in-house information unless otherwise indicated. Prices include installation and taxes but do not include the available financial incentives in each country.

51   The LUEC was calculated using a 20-year life and a 10 per cent real discount rate. The corresponding nominal discount rate would be on the order of 8 per cent.

52   Stenkjaer, N., 'Solar thermal collector', *Nordic Folkecenter for Renewable Energy*, www.folkecenter.net/gb/rd/solar-energy/solar_collectors.

53   Building Research Institute, 'Reducing CO2 by low-cost solar hot water system located in condominium veranda', *Japan for Sustainability*, 2008, www.japanfs.org/en/pages/028584.html.

3
Review of Programmes

As shown in Exhibit 4, the next project step was to select and review successful programmes. This section discusses:

- Selection of programmes for further investigation.
- Data collection and programme summary development.
- Programme trends.
- Definitions, drivers and indicators of success.

3.1 Selection of programmes for further investigation

As discussed in Section 1.4.2, programmes were selected for further investigation based on the following criteria:

- Evidence that programme/policy likely caused a significant increase in renewable technology uptake based on either:
  - programme evaluation reports; or
  - country or region has either high installed capacity or a recent high rate of installed capacity under circumstances that are not all favourable;
- Programme has been unusually stable and continuous over a period of time.
- Programme has unique features not found in other programmes/policies being studied.
- Enough historical data is available to allow a determination of how well the programme is working and what lessons have been learned.
- Programme of sufficient scale (time, region covered or budget) to have made an impact.
- Selection from the most successful countries, as determined in Chapter 2.
- Programme provides diversity among programmes/policies chosen, in terms of:
  - technology(ies) targeted;
  - type of programme;
  - dwelling types and markets targeted.

A long list of candidate policies and programmes was generated from the country review presented in Chapter 2. In addition to the country-specific programmes, a few multi-country efforts identified during the country review were included on the list. Exhibit 14 in Chapter 7 presents the candidate list and indicates whether these programmes meet the first five above-listed criteria.

Once programmes were chosen using the above criteria, a check was made to ensure that all types of programme instruments, all of the target technologies and all sectors were included. These analyses are presented in Exhibits 15–17 in Chapter 7 and indicate that the chosen programmes adequately cover the instrument types, technologies sectors.

3.2 Data collection and development of programme summaries

In-depth programme information was collected from a combination of publicly-available information sources and personal interviews with programme personnel. The detailed programme summaries are available in Chapter 7.

3.3 Trends and patterns in successful programmes

After the detailed programme information was collected, but before selecting best practices, we examined the successful programmes for trends and patterns. We observed the following:

- Austria and Germany are recognized as leaders in per capita solar installations, in spite of their modest solar resources and cool climates. One of the factors in this success may be the long history of solar programmes in those countries. In the case of Austria, programmes have been continuously funded for 30 years in some regions. The integrated Market Incentive Programme (MAP) in Germany has been funded for ten years. Long-term continuous programmes appear to be one factor promoting success.
- Many of the successful programmes studied were multifaceted. They combined an economic incentive 'carrot' with a 'guidance' educational campaign. In some cases, the educational campaign had several components. For example, one part of the campaign might target the general public, while another part of the campaign targeted installers. Finally, several of these programmes also include regulatory 'sticks' that mandate the use of renewable thermal energy under certain circumstances. Other studies, such as the World Energy Council's 2008 *Energy Efficiency Policies Around the World: Review and Evaluation* report, have observed similar patterns. This report states: 'Packages of complementary measures [are] an effective way to speed up the development of new technologies, such as solar water heaters.'
- A recent trend observed in many of the studied countries is the move to requiring the use of renewable thermal energy in new buildings. Some countries also require that renewable heating technologies be installed in

buildings undergoing major renovations. Regions of Austria and Italy, and Spain now require minimum amounts of solar heating in certain new homes. Germany and Italy require the use of minimum amounts of renewable thermal energy in new homes.

3.4 Definitions, drivers and indicators of 'success'

A review of the programmes also revealed that a variety of definitions of success are being used by programme managers and funders. This indicates that programmes cannot be compared without considering the possibility that they have been developed to reach achieve different goals. Exhibit 13 lists some of these definitions, drivers (objectives or interests) and indicators.

Definition	Driver	Indicator(s)
Increased deployment of technology	• Reduce conventional energy use (for environmental or security reasons). • Reduce greenhouse gas emissions. • Address electricity distribution issues (in areas where electricity is used for heating). • Avoid extending conventional fuel networks.	• Additional capacity installed. • Proportion of households with technology installed. • Number of applications completed.
Improved quality of installations	• Address installation quality issues in a region where this has been a problem. • Run a small-budget programme (funding insufficient for large-scale subsidies).	• Customer satisfaction with installation. • Number installations done with certified equipment. • Number of installations that exceed minimum equipment standards. • Change in quality of equipment available in the local market, or change in the level of training or experience of installers.
Increased deployment of technology in niche markets not reached by other programmes	• Reduce conventional energy use. • Reduce greenhouse gas emissions. • Address electricity distribution issues (in areas where electricity is used for heating). • Avoid extending conventional fuel networks.	• Additional capacity installed. • Number of applications completed.

Exhibit 13 Sample definitions, drivers and indicators of success

4

Best Practices

As shown in Exhibit 4, the next step after the detailed programme review was to study both the country profiles and the detailed programme profiles to identify 'best practices'. The selection criteria for these practices are presented in Section 4.1.

These best practices were then organized according to the framework described in Section 4.2. The resulting best practices guide is available in Part 2.

Finally, the best practices were analysed. These findings can be found in Section 4.3.

4.1 Selection of best practices

A practice or lesson learned was considered a 'best practice' if:

- It was found to be widely used in successful programmes.
- It was found to be used, often more than once, in countries that have been particularly successful at promoting renewable technologies.
- It was found in a programme that has been particularly successful and appears to have been essential to that programme's success.

4.2 Best practice guide organization

The best practices were organized according to programme phase, market stage and barriers addressed. We discuss each of these categories below.

4.2.1 Phases of programme development and associated needs

As discussed in Section 1.3.2, REHC programme development can be broken into four phases: portfolio planning, programme design, programme implementation and programme evaluation. Programme needs vary for each of these programme phases:

Portfolio planning: During this phase, portfolio planners focus on understanding the technical, economic and achievable potentials of REHC technologies. They also investigate the barriers to greater use of REHC technologies in the target areas, identify and learn about the interests of the stakeholders and attempt to determine the decision-making and key leverage points. Typically a market study is done during this stage. Once portfolio planners collect and analyse this information, they consider the various programme possibilities and evaluate the suitability of each. Portfolio planners aim to establish clear, consistent, ambitious but achievable goals and develop programmes and plans to achieve them.

Programme design: During this phase, programme designers collect information on best practices and experiences from other programmes, particularly from areas similar to the target area. If a market study was not done during the portfolio planning phase, it may be done during this phase. Programme designers attempt to determine the best programme type, an optimal level of incentive or regulation and methods of minimizing free ridership. They develop relationships with potential partner. They plan for implementation, evaluation and scheduled modifications of the programme.

Programme implementation: During this phase, programme implementers implement the programme as designed. They watch for problems and monitor results. Programme implementers typically make small programme changes as needed, leaving larger programme changes for scheduled modification points.

Programme evaluation: During this phase, programme evaluators review the programme logic, establish evaluation objectives (such as relevance, results and cost-effectiveness), collect and analyse data and feed the results back to the personnel from other programme stages so programme modifications can be made and impacts can be determined.

Best practices are often, but not always, specific to programme phase.

4.2.2 Stages of market development and associated needs

Section 1.3.3 describes how programmes designed to develop REHC markets can be categorized according to the three types of markets: 'initial deployment', 'mass market' and 'full market'. Programme needs vary for each of these market stages:

Initial deployment: During the initial deployment stage of a technology in a given market, programmes need to demonstrate that the technology is beneficial and reliable, encourage small groups of people to adopt the technology and start to build the infrastructure required to supply, install and support the technology. Some consumers will be ready to adopt the technology more quickly than others and may be motivated by non-economic factors such as environmental concerns. During this stage, the technology may be more economically attractive in certain niche markets.

Mass market: During the mass market stage of market development, programmes must encourage larger numbers of the public to adopt the technology. This is most often done with a combination of 'carrot' and 'guidance' measures, because consumers will be concerned about cost and may not understand the technology or the benefits. Programmes also need to further develop the market infrastructure in preparation for the full market stage.

Full market: During this market stage, technology use has become widespread and the market infrastructure is fully developed. Programmes at this market stage need to use 'sticks' to motivate consumers who are focused on initial rather than life cycle costs. 'Carrots' and 'guidance' measures may also still be appropriate to level the market difference between conventional and renewable technologies.

While some best practices are applicable to all three market stages, many are specific to one or two.

4.2.3 Barriers and best practices that address these barriers

REHC technologies face a variety of barriers to wide-spread use. Section 1.3.4 describes the five 'A's method of categorizing these barriers (acceptability, accessibility, affordability, availability and awareness).

A full barrier analysis was not included in this project, however, to facilitate the identification of the most appropriate best practices for a given situation, barriers and their corresponding best practices were identified.

4.3 Best practice analysis

An analysis of the best practices collected during this project leads to the following observations:

- No central sources of best practice information.
- Technology type is less important than programme phase, market maturity and barrier faced.
- Identified best practices address all categories of market barriers but not all barriers.
- Some best practices are both globally applicable and essential.
- Other best practices are situation-specific.
- There is a need for more evaluation.

These observations are discussed below.

4.3.1 No central sources of best practice information

During the course of our research, we did not find centralized sources of best practice information about renewable heating and cooling support programmes in the residential sector. Some of the programmes we reviewed had knowingly incorporated elements of other successful programmes, however, there do not appear to be formal mechanisms for sharing best practice information.

4.3.2 Technology type is less important than programme phase, market maturity and barrier faced

Most of the identified best practices are not technology-specific but rather are specific to programme phase, market stage and barrier faced. This suggests that the body of experience developed with energy conservation programmes may also be applicable to programmes targeting REHC technologies. Best practices developed for solar photo-voltaic support may also be applicable.

The one target group of technologies that requires an additional consideration is energy-efficient biomass stoves or boilers. Unlike other renewable technologies, biomass installations require a fully-developed fuel supply chain involving manufactures, distributors and retailers. Wood pellets, for example, must be reliably and conveniently available at a predictable price if pellet-burning stoves are to be acceptable.

4.3.3 Identified best practices address all categories of market barriers but not all barriers

Of the 32 best practices identified, two-thirds address the five market barriers noted in Section 4.2.3, with fairly even coverage. (The remaining best practices address programme-specific barriers, such as the sponsoring agency not having enough programme design and implementation expertise on staff.)

Although all categories are covered, not all specific barriers within each category are covered. Since the study did not include a thorough barriers analysis, this finding is based on anecdotal observation. For example:

- **Acceptability:** None of the best practices addresses the issue of potential reduced convenience of operation of REHC systems (for example, taking showers in the evening instead of in the morning).
- **Accessibility:** None of the best practices addresses regulations that may prevent or control the installation of biomass combustion devices.
- **Affordability:** None of the best practices addresses methods of reducing technology costs, such as bulk buying cooperatives or promoting solar-ready homes.
- **Availability:** None of the best practices addresses the concern that there may be little consumer choice in some niche markets.
- **Awareness:** None of the best practices addresses the concern that some categories of consumers may uninterested in typical information packages or unreachable through traditional promotion campaigns.

This suggests that, beyond the adoption of best practices documented in this study, the promotion of REHC would benefit from a top-down barriers analysis and a systematic consideration of potential innovative new approaches to overcome barriers.

4.3.4 Some best practices are both globally applicable and essential

Some of the identified best practices appear to be globally applicable to all programme types and essential to success. These include:

- **Set clear, consistent, ambitious but achievable targets.** Clear targets are essential for good programme design.
- **Establish plans and programmes to achieve those targets.** These plans and programmes should address all market participants to avoid 'slow' participants from blocking overall market progress.
- **For each programme, establish clear objectives, targets and performance indicators.** These parameters facilitate programme design, evaluation and refinement.
- **Monitor and report progress towards the targets.** This ensures accountability and allows for periodic programme corrections.
- **Design and implement evaluations to assess success, then refine targets and programmes.** Without evaluation, it is difficult to determine success and to make corrections.

In many cases, the following are also both applicable to all programme types and essential:

- **Establish market needs and barriers prior to programme design.** Successful and effective programmes address the needs and barriers of a particular market. Unless the needs and barriers are already known, a market study is essential.
- **Design and implement multifaceted programmes.** In almost all situations, a successful programme will need to address multiple groups of market participants in multiple ways. For example, a training programme for installers might address a lack of experienced installers while an incentive programme might encourage individual households to install a new technology.
- **Design programmes to complement, not conflict, with other programmes in the region.** Programmes that conflict will, by definition, make one of the programmes less successful. On the other hand, complementary programmes are likely to be more effective than one programme alone because they will collectively address more of the market barriers and needs.
- **Design programmes with adequate flexibility to adapt to market changes.** Market change is inevitable, unless a programme is very short-lived. Technology costs will change, new technologies will become available, other programmes will start up or end, or conventional energy prices will

change. In order to be successful, programmes must adapt to accommodate these changes.

- **Ensure that the market infrastructure can successfully deliver and install products being promoted.** Subsidies, requirements and educational efforts will be unsuccessful if renewable energy technologies can't be delivered and installed. Market development measures may need to precede or complement measures aimed at consumers.
- **Include entry and exit strategies in programme design to avoid undesirable market distortion.** Poorly designed or implemented programmes can cause price or availability distortions, threatening future programme success. For longer–term programmes, possible market distortions must be considered when planning and implementing periodic programme adjustments.

4.3.5 Other best practices are situation-specific

In contrast to the more globally-applicable best practices listed above, other best practices are specific to the situation at hand. These include:

- **Consider having a third party design, implement and/or evaluate the programme.** Independent evaluation is always a good practice. The appropriateness of third-party design and implementation will depend on the expertise of the sponsoring agency.
- **Break longer term targets into shorter-term milestone targets.** In some instances, both the sponsoring party and the public may find shorter-term milestone targets easier to work with.
- **Design flexible plans and tools to support regional or municipal progress towards the established targets.** In some cases, lower levels of government may be responsible for implementing programmes that are part of a larger effort. Higher levels of government, or non-governmental bodies, may be able to develop plans and tools that facilitate this process.
- **Identify and encourage 'early adopters'.** Some individuals or organizations are more willing to adopt new technologies or processes. Identifying and encouraging these parties can move less enthusiastic parties to action.
- **Design and implement long-term programmes.** With very few exceptions, the successful programmes are long-term.
- **Penalize non-compliance with 'stick' programmes.** Requirements, such as for installation of a certain amount of renewable energy in new homes, need to be backed by consequences to be effective.
- **Develop and run programmes in close cooperation with relevant public bodies and industry partners or organizations.** Many of the successful programmes studied encouraged the involvement of multiple market players.

- **Engage media to provide publicity.** Developing partnerships with both general and industry-specific media can help provide the publicity that is often essential to programme success.
- **For information campaigns, invest in high-quality promotion materials.** Well-written and attractive materials encourage the end-consumer to participate.
- **Ensure a central point for accurate information.** If market participants might get confused about technical or programme details, establishing a single point of information will reduce misinformation.
- **Get committed renewable energy champions to participate.** Some successful programmes have found that renewable energy champions, including politicians, can be important programme partners.
- **Include an evaluation component in the programme design.** Programme evaluation directs programme corrections either for the current programme or for future programmes.
- **Limit information campaigns to specific, clear messages.** Addressing too many topics, barriers or market participants at the same time is confusing.
- **Keep business interests in the background during information campaigns.** Renewable heating and cooling businesses can be very helpful partners during information campaigns, however, the public will be most receptive when the information is presented by independent experts.
- **Develop and distribute tools that facilitate household participation.** Heating cost calculators and installer databases are two types of tools that help households decide to participate in programmes.

The following best practices also fall into this 'situation-specific' category but are applicable only to incentive programmes:

- **Consider offering additional incentive amounts for higher quality equipment or additional efficiency measures.** When installation quality is a problem, 'bonus' incentives for high-quality installations may remedy the problem. Additional incentives can also motivate people to install more than the minimum system.
- **Consider requesting technical drawings from installers.** Requesting and reviewing technical drawings prior to paying an incentive is another possible solution when installation quality is a problem.
- **Regulate the quality of the equipment and installations funded by incentives.** Requiring evidence of quality helps ensure that installations perform as expected. Installation of poor systems will discourage the market.
- **Offer incentives by energy (kWh) or capacity (kW or m²) rather than as a percentage of cost.** Market price distortions may occur when incentives are calculated as a percentage of the installed cost. Incentives

based on energy produced or capacity are less likely to result in oversized or overpriced installations.

- **Consider requesting an energy audit certificate with the incentive application.** This will be particularly important if the installation of oversized equipment has been a problem or if implementation of energy efficiency measures is also being encouraged.
- **Design application process to be efficient for both homeowners and programme administration.** A complicated application process will both discourage homeowners and increase programme administration costs.

Need for more programme evaluation

A number of evaluation best practices were observed but the overall impression left by the programme case studies is that evaluation practice is not well-developed in the context of renewable heating and cooling. In part, this is because a number of the programmes studied have not been formally evaluated. As a result, it is difficult to conclude, in some cases, whether the reported results are real and accurate, and whether they are truly attributable to the programme, or to other external factors such as variations in the price of conventional fuels.

This stands in contrast with current approaches in energy efficiency, where programme evaluation has taken a central role in providing credibility for reported results and identifying opportunities for improved programme portfolios, programme design and implementation. This is particularly true in the context of utility demand-side management (DSM) efforts but also applies to a variety of energy efficiency programmes. This body of work has largely been driven by the methods and approaches used to evaluate North American utility programmes where state and provincial regulatory oversight demands very high levels of rigour to the evaluation work. Three key sources of methodological guidance are the *Energy Efficiency Policy Manual*, the *California Evaluation Framework*, *California Energy Efficiency Evaluation Protocols* and the *International Performance Measurement and Verification Protocol (IPMVP)*:

- The California Public Utility Commission (CPUC) *Energy Efficiency Policy Manual* directs programme evaluation scope and approaches and the *California Evaluation Framework* provides a consistent, systemized, cyclic approach for planning and conducting evaluations.[1] The *California Energy Efficiency Evaluation Protocols*[2] provide more detailed guidance.
- The IPMVP is the key guidance document for the measurement and verification of energy savings.[3] A companion volume also deals with renewable energy applications: *IPMVP Volume III: Part II Concepts and Practices for Determining Energy Savings in Renewable Energy Technologies Applications*.

Though not represented in the case studies, there appears to be an opportunity to improve the practices in renewable heating and cooling programme evaluation by adopting some of the best practices used in the context of energy efficiency.

Notes

1   TecMarket Works et al, *The California Evaluation Framework* (California Public Utilities Commission, June 2004). Available at www.cee1.org/eval/CEF.pdf.
2   TecMarket Works Team, *California Energy Efficiency Evaluation Protocols*, California Public Utilities Commission, 2006. Available at www.calmac.org/events/EvaluatorsProtocols_Final_AdoptedviaRuling_06-19-2006.pdf.
3   Efficiency Valuation Organization – International Performance Measurement and Verification Protocol Committee, *International Performance Measurement and Verification Protocol*, California Public Utilities Commission, March 2002. Available at www.nrel.gov/docs/fy02osti/31505.pdf.

5
Conclusions and Recommendations

5.1 Review of country experience with REHC technologies

The use of REHC technologies varies considerably by country. In some cases, external factors can explain high REHC use. In other cases, policies and programmes appear to favour REHC installation even when the external factors, such as renewable resource availability, do not.

Installed solar capacity has been well-studied, facilitating comparisons between countries. Among the group of countries studied, Austria was the clear leader in per capita total installed solar capacity. Denmark, Germany and Japan were ahead of the remaining countries, but were significantly below Austria. Interestingly, Austria, Denmark and Germany have relatively low annual solar insolation levels, and both Austria and Germany have high relative technology costs.

Small-scale energy-efficient biomass technologies have been less well studied. Some information on the installed number of pellet burners and on annual quantities of pellet productions is available. However, because other energy-efficient biomass technologies exist, and because a large percentage of pellets are exported, it is impossible to determine which countries are leaders in this area.

Similarly, it is difficult to find comparable information on air-source heat pump installations by country. This may be because air-source heat pumps are not considered renewable or alternative technologies in some areas and therefore have not been comprehensively studied.

Information on the number of ground source heat pumps installed per country is only marginally more available and is often contradictory. Among the countries studied, and where data was available, Austria and Norway appear to be per capita ground source heat pump leaders.

Levelized unit energy costs of thermal solar installations in the US, Denmark and Japan showed that Japan and the US had the most competitive costs, whereas the costs were significantly higher in Denmark. However, even in Japan, the cost of solar heating exceeded the cost of electricity and natural gas. This suggests that solar heating continues to rely on financial incentives to overcome financial barriers in all countries.

5.2 Review of successful REHC programmes

Many interesting and successful REHC support programmes were identified in the countries studied during this project. Some of the most successful programmes are long-term, as exemplified by the long-standing solar support programmes in Austria and Germany. Many of the successful programmes were also multifaceted, addressing multiple audiences with a variety of support instruments. The Upper Austria Energy Action Plan, for example, recognizes that slow development of one group of market participants can become a bottleneck for the overall development of a renewable market.

A trend observed in many of the studied countries is the move to requiring the use of renewable thermal energy in new or renovated buildings. These may be solar-specific, as found in some regions of Austria and Italy, and in Spain; or may be general renewable energy requirements, as found in Germany and Italy. In most cases, these requirements have been preceded by a long period of voluntary participation in information and incentive programmes. The voluntary programmes appear to have fostered citizen support for mandatory requirements and a mature renewable market capable of supplying the technologies.

REHC support programmes have a surprising variety of objectives. While some are focused on increasing the general technology uptake, others focus on improving quality or reaching niche markets. Caution should be used when comparing programmes with differing objectives.

5.3 Best practices for REHC programmes

The studied programmes are rich in best practices and lessons learned. For the most part, these are not technology-specific but rather are linked to programme phase, market stage and/or barrier faced. This suggests that the large body of experience developed in the delivery of energy conservation programmes or photo-voltaic solar support programmes may also be applicable to programmes targeting REHC technologies. The one target group of technologies that requires additional consideration is energy-efficient biomass stoves or boilers, because of the need for a well-developed and reliable fuel supply chain.

Some of the identified best practices are globally applicable to all programme types and are essential to success. These include: setting clear, consistent, ambitious but achievable targets; establishing plans and programmes to achieve those targets; monitoring and reporting progress towards the targets; and designing and implementing evaluations to assess success, then refining targets and programmes. A number of additional best practices are applicable to most programme types and also appear to be essential to success. Examples of this type of best practice are: establish market needs and barriers prior to programme design; and design and implement multifaceted programmes. The remaining best practices are situation-specific.

Specific best practices identified during this project have been summarized in Part 2. This guide is designed to provide portfolio planners and programme designers, implementers and evaluators with concrete advice based on the review of successful REHC programmes.

The overall impression left by the programme case studies is that REHC programmes have not been sufficiently evaluated. This makes it difficult to verify conclusions about the success of a programme, and also may mean that programmes are not as effective and efficient as they could be. Evaluation of energy efficiency support programmes has been more robust, particularly in the context of utility demand-side management (DSM) programmes in North America. Evaluation methods and tools developed for energy efficiency programmes should be considered for REHC support programmes.

5.4 Key conclusions

The key conclusions of this study are:

- **Appropriate best practices depend on the context.** This includes the objectives (for example, increasing deployment, improving quality, filling niches, cost-effectiveness, and so on), the authority involved and the programme development phase.
- **Appropriate best practices depend on the level of market maturity.** This means that it is important for policymakers to appreciate the level of penetration of the various technologies.
- **Best practices need to respond to the relevant barriers.** This means that policymakers need to understand the barriers faced by REHC, including both market barriers of acceptance, accessibility, affordability, availability and awareness, but also practical barriers such as lack of information, lack of capacity and lack of consistent political direction.
- **There is a wealth of experience with best practices available to be shared with policymakers.** These address many of the potential barriers and cover all barrier types. They have been captured and documented in Part 2.
- **Some barriers have not been sufficiently addressed.** Although documented best practices address many barriers, they do not address all of them. This means that there is a need for further research to understand barriers, and a need for ongoing policy innovation. Fortunately, we found evidence that this innovation is happening and identified several new programmes that were too recent to be included in the scope of the present study. We also identified other areas of energy policy (for example, energy efficiency, photo-voltaic support) that face similar barriers and that could be investigated for more best practices.

- **The effectiveness of best practices is unknown.** The best practices included in Part 2 were mostly chosen on the basis of subjective evidence of success. The reality is that most REHC programmes are not being evaluated with any reasonable degree of rigour. This means that the outcomes of the programmes (and, by extension, the associated best practices) are not being adequately measured, verified and reported.
- **Awareness of REHC best practices is uneven.** Good sources of best practice information and mechanisms for sharing information about best practices and lessons learned in other jurisdictions have not been developed.

5.5 Recommendations

Based on the above conclusions, we offer the following recommendations:

- **Policymakers should begin by developing a consensus understanding of the context for the intervention.** This means understanding and clearly articulating the problem and the REHC objectives, as well as appreciating the limits of the relevant authority and the maturity of the policy development (i.e. phase).
- **Before proposing REHC programmes, policymakers should invest in market research.** They need to understand the relative penetration of the various technologies (i.e. market maturity), as well as the barriers to market penetration (for example, in terms of the five 'A's).
- **Policymakers should make use of the companion *Renewable Heating and Cooling Best Practices Guide* in Part 2.** By consulting the guide, they can find information on relevant approaches that have been used to overcome similar barriers in comparable situations.
- **Policymakers should try new approaches.** Beyond the examples in the guide, there will be additional barriers and circumstances that require different practices. Some may be borrowed from other areas of energy policy or adapted from academic and public policy research. This project focused on the bottom-up collection of best practices from real programmes, as opposed to the systematic examination of barriers and solutions. Another study commissioned by IEA-RETD (*RENBAR – Good Practices for Solving Environmental, Administrative and Socio-economic Barriers in the Deployment of Renewable Energy Systems*) is addressing barriers more systematically but is not focused on REHC technologies. It is recommended that IEA-RETD explicitly address the links between the two studies and fill the gaps in the *Renewable Heating and Cooling Best Practices Guide* by incorporating applicable guidance for barriers that were not addressed in this REHC project. IEA-RETD should also consider follow-up research to extract best practices from the newer, more innovative programmes identified during this study.

- **Policymakers should conduct robust evaluations of their programmes.** Governments and stakeholders need credible information on the relevance, success and cost-effectiveness of REHC programmes in order to determine whether they represent appropriate investments of public resources and to determine which best practices truly contribute to success. This means adopting modern evaluation methods regarding measurement and verification, as well as dealing with attribution issues such as free ridership and spillover.
- **IEA-RETD should promote greater awareness of REHC best practices.** This includes promoting the *Renewable Heating and Cooling Best Practices Guide,* as well as monitoring its use and updating its content. It should also include promoting networks of REHC policymakers and sponsoring events that would enable such policymakers to share best practices and lessons learned. IEA-RETD should identify other regional or international organizations with similar objectives to share the burden and leverage additional intellectual capacity.

5.6 Additional recommendations for future work

The following additional research is suggested by the current work:

- The lack of comparable, systematically-collected information on residential energy-efficient biomass and heat pump installations makes it difficult to determine which regions are most successfully promoting these technologies. Consideration should be given to the collection of this information.
- During the course of the project, a number of innovative RHEC programmes were identified (see the 'Innovative programmes' section of Chapter 7). These programmes are currently too new to provide best practices, but should be reviewed in a few years for additional best practices.
- As mentioned above, best practices developed for energy efficiency and solar photo-voltaic programmes may be applicable to REHC programmes. This possibility should be explored further.
- This project focused on small- to medium-scale renewable thermal installations. Large-scale renewable thermal installations were not studied because they present different challenges and opportunities. A project focusing on best practices for these installations could provide guidance to programme developers, implementers and evaluators of larger-scale installations.
- In order to limit the scope of this project to a manageable set of tasks, the country reviews were limited to high-level observations. In some cases, important cause and effect relations will have been missed. (For example, regions with programmes focused on quality may have had significant earlier problems with programmes that inadvertently promoted poor-quality

systems.) The Intelligent Energy Europe Res-H project is compiling more in-depth country profiles for seven countries/regions: Upper Austria; Austria, including Styria; Greece; Lithuania; the Netherlands; Poland and the UK. Performing similar in-depth country profiles for the remainder of the IEA-RETD countries could be considered.

- This project focused on the bottom-up collection of best practices from real programmes, as opposed to the systematic examination of barriers and solutions. Another study commissioned by IEA-RETD (*RENBAR – Good Practices for Solving Environmental, Administrative and Socio-economic Barriers in the Deployment of Renewable Energy Systems*) is addressing barriers more systematically but is not focused on REHC technologies. It is recommended that IEA-RETD explicitly address the links between the two studies and fill the gaps in the *Renewable Heating and Cooling Best Practices Guide* by incorporating applicable guidance for barriers that were not addressed in this REHC project.

6
Country Summaries

6.1 Austria

Austria has historically offered a variety of incentives for residential renewable heating. Natural gas and electricity are taxed, in part, to fund alternative energy programmes. Solar programmes have been continuously funded throughout the last 30 years in some parts of Austria. Incentives for geothermal and biomass energy are also catching up.

Context

The Republic of Austria is a federal republic divided into nine states (*Bundesländer*). Pertinent energy legislation and administration (such as building codes and renewable energy subsidies) are primarily the responsibility of the states. The federal government sets the legal framework and the states implement the laws with their own rules. Electricity is produced from hydropower, combustion (coal, gas and biomass) and by cogeneration. No nuclear power is produced. A smaller portion of electricity is imported.[1] Natural gas and heating oil are widely available throughout the country. Natural gas is largely imported from Russia, Norway and Germany.[2] Crude oil is imported from Saudi Arabia, Libya, Kazakhstan and the Russian Federation.[3] Approximate conventional energy prices are presented in Exhibit 5. The country has relatively low access to solar resources and good access to biomass and low-temperature geothermal resources. The climate ranges from temperate to alpine, with cold winters and hot summers.

Technology market status

Good quality solar thermal systems for residential applications are readily available at a cost of approximately €820–1150/kWh including taxes.[4] In 2005, 500,000 tile stoves were installed in Austria[5] for use either as sole or additional heating systems. These primarily burn logs. Pellet-fired tile stoves and low capacity tile stoves have been introduced to the market. Pellet or wood chip boilers are gaining popularity with almost 50,000 pellet boilers installed in 2006.[6] Pellet supply and prices have been volatile. Estimates of heat pump installations vary widely: one source suggests that by the end of 2006, 200,000 heat pumps had been installed. Most of these were ground source heat pumps (70 per cent).7 Another source suggests that the number of ground source heat pumps in use in 2007 is on the order of 40,500.[8]

Programmes

The main issues of the Austrian energy policy are sustainability and security of supply. In order to increase the security of supply and reduce the energy import dependency, the Austrian government plans to increase the use of renewable energy resources, domestic energy production, and the research and use of energy conservation and efficiency technologies.[9]

While respecting the European Union (EU) objective of 20 per cent renewable energy by 2020, the specific target for total energy from renewable energy sources in Austria is 34 per cent by 2020. (By 2005, 23.3 per cent had already been achieved.[10]) Many villages and cities, and eight out of nine states, have joined the Climate Alliance of European Cities with Indigenous People, with its aim of a 50 per cent reduction of CO_2 emissions by 2010.

To accomplish the above goals, Austria has relied on a variety of programme types: financial incentives, market development initiatives, training for installers and voluntary labelling programmes. The programmes are administered primarily at the state and municipal level, rather than at a federal level. Each state generally has its own specifications for their subsidies, and they have attained varied, but generally good, results.

Results

Austria has achieved consistent market growth through subsidies and requirements for solar thermal technologies, resulting in a very high solar thermal capacity per capita. Biomass heating has experienced similar success. The geothermal heat pump market has also grown steadily, although not as quickly as for biomass and solar heating, as a result of subsidies offered in all states but Steiermark. Heat pumps are the most successful heating system in new buildings, however, slow growth has resulted from the fact that they are less often used when retrofitting existing buildings.

Significant programmes of interest

- *Klima: aktiv* programme campaign included the *Wärmepumpe* promotion for heat pumps, the *Holzwärme* promotion for biomass heating and the *Solarwärme* promotion for solar thermal heating.
- Energy Action Plan Upper Austria.
- Regional promotion programmes, including *Spar mit Solar*, *Ja Zu Solar*, and the Salzburg and Vrarlberg regional subsidy programmes.
- Residential Solar Obligation.

6.2 Canada

Canada has a wealth of resources but lacks a unified policy for the promotion of renewable heating and cooling applications. Biomass along with solar-thermal represent the greatest potential resources for renewable energy for heating, although more must be done to develop the supply chains and to inform the general public of their benefits.

Context

Canada is composed of ten provinces and three territories. Energy policy is primarily administered at provincial/territorial level. Canada produces oil, natural gas, hydroelectricity, nuclear power (powered by uranium also available in Canada) and coal. In general, Canada exports significantly more energy than it imports.[11] Natural gas is widely available in most urban areas but has, until recently, been less available in the Atlantic Provinces. Approximate conventional energy prices are presented in Exhibit 5. The country has good access to solar resources, good access to biomass resources and variable access to low-temperature geothermal resources. The climate varies from temperate in south with cold winters and warm summers, to sub-Arctic and Arctic in the north.

Technology market status

Good quality solar thermal systems for residential applications are readily available at a cost of approximately €4000 installed including taxes.[12] Information on the number of energy-efficient biomass heating installations is not available. The domestic supply chain for pellets is not well-developed. In 2007, 1.4 million tonnes of pellets was estimated for production, of which all but 200,000 tonnes were exported.[13] Total number of installed air-source heat pumps is not available, but one source estimates that approximately 31,000 ground source heat pumps were in use in 2007.[14] While some areas of the country have climates well-suited to air-source heat pumps, most areas require supplemental heating during the winter. Cold-climate air-source heat pumps are under development. Ground source heat pumps are better suited to the climate in many areas, but also require supplemental winter heat if they are sized correctly for summer use.

Programmes

The main principles of energy policy in Canada are market orientation, respect for jurisdictional authority and provincial governance, and targeted intervention in the market process to further specific policy objectives. Although no explicit federal policy on renewable energy exists, greenhouse gas emission reduction targets are 17 per cent from 2005 levels by 2020.[15] Climate change policy has been moving forward at the provincial level in several provinces including British Columbia and Ontario.

The main mechanisms Canada is using to further these policies are financial incentives for energy efficiency. Renewable energy programmes also provide subsidies at the federal and provincial level.

Results

Heating from renewable sources has experienced limited market growth in Canada. Solar thermal heating has yet to impact Canada in any significant way. Energy-efficient biomass use is limited despite the large quantity of biomass available. The pellet industry production is high, but most of the pellets produced are exported and the supply chain for fuelling domestic demand within the country is in need of development. Despite receiving limited support from incentives, heat pumps have experienced a moderate amount of market growth in all but a few provinces.

Significant programmes of interest

- Renewable Energy Deployment Initiative (REDI).[16]
- Manitoba Residential Earth Power Loan.

6.3 Denmark

Denmark has realized significant progress in energy efficiency and renewable energy through carbon and energy taxes. In particular, great gains have been made in district heating, some of which is renewable. Regulations, mostly for biomass and solar thermal heating, have been particularly effective in growing the renewable heat market.

Context

Denmark is a constitutional monarchy divided into five regions (*regioner*). Municipalities are given energy-planning tools from a centralized federal energy authority.[17] Denmark has sizeable sources of oil and natural gas in the North Sea, which it exports, while also importing solid fuels.[18] Although oil and imported coal are the most important fuels for electricity generation, natural gas and renewable resources have been slowly replacing these fuels. Conventional energy prices are presented in Exhibit 5. The country has relatively low access to solar, moderate access to biomass resources and good access to low-temperature geothermal resource. Denmark has a temperate climate, with an average of 121 days of precipitation, mostly in the autumn. Its northern location means great variations in daylight hours throughout the year. Prices of conventional energy sources are high.

Technology market status

Good quality solar thermal systems for residential applications are readily available at a cost of approximately €4000–4700 installed including taxes.[19] In 2006, 500,000 wood-burning stoves, 70,000 wood-burning boilers, 30,000 wood pellet furnaces and 9000 straw-burning furnaces were installed.[20] One source states that 40,000 households were heated with heat pumps in 2006.[21] Another suggests that a total of 11,250 ground source heat pumps were in use in 2007.[22]

Programmes

In the Energy Policy Statement 2008,[23] Denmark indicated that efforts should be put towards energy savings and enhancement of energy efficiency, increased use of renewable energy, and technological development. Current national targets include the reduction of total energy consumption by 2 per cent in 2011 and by 4 per cent in 2020 (compared to 2006 consumption), and the increased use of renewable energy to 30 per cent of final energy consumption by 2020 (20 per cent by 2011). Until 2001, several mandates for connection to district heating grids and bans on electric heating appliances greatly increased renewable resource utilization for heating. In 2001, support was disrupted by the liberalization of the electricity market that resulted from a shift of political power.

Denmark has had great success with heating energy policy and programmes. The first effort was the Heat Supply Law in 1979 as part of the Danish Energy Policy which required local authorities to report on heating requirements, methods and energy consumed for their community. Energy Plan 81 was drafted in 1981 to further support heat and electricity generation initiatives.[24] Now, more than half of the households in Denmark use district heating. To support the policies and targets, Denmark uses financial incentives (for example, the Development Programme for Renewable Energy), as well as regulations and codes (such as the Executive Order on Connection to Public Heat Supply Installations and building codes).

Results

Denmark has a relatively low installed capacity of solar heating per capita and a low rate of increase. The air-source heat pump market in Denmark is experiencing significant growth whereas the ground-source heat pump market development has been moderate. Denmark has low biomass resources and limited biomass market development.

Significant programmes of interest

- Executive Order Solar Heating Obligations.
- Biomass Agreement.

6.4 France

Although activity in France lags behind that of some of its neighbours, France is starting to further encourage renewable heating and cooling through regulations and tax credits. The geothermal heat pump market is currently growing.

Context

France is a republic divided into 26 administrative regions, 22 of which are located in Europe and four of which are overseas. The regions have limited authority in terms of governance and energy policy. France has very limited energy reserves; the country is heavily reliant on nuclear energy for energy production and also relies on imported oil and gas. Conventional energy costs are presented in Exhibit 5. The country has moderate access to solar and biomass resources and good access to low-temperature geothermal resources. France is located in a northern temperate zone. The north and northwest have a temperate climate with maritime influences, while a Mediterranean climate prevails in the south.

Technology market status

Good quality solar thermal systems for residential applications are readily available at a cost of approximately €1695, including tax but not including installation.[25] Most of the fuel wood is used in stoves and fireplace inserts. In 2006, only 28,000 biomass-fired boilers were sold.[26] In 2007, 200,000 tonnes of pellets were produced, 165,000 of which were used in France.[27] Also in 2007, 51,000 air source heat pumps[28] and 18,600 ground source heat pumps were sold.[29]

Programmes

The aim of French energy policy is to limit the exposure of the economy to fluctuations in energy prices by developing domestic production. The government supports controlling energy demand, diversifying sources of energy and providing methods of energy transport and storage. Current renewable energy policies follow EU targets, including 21 per cent electricity from renewable sources by 2010 (mechanisms include feed-in tariffs and a tender system for large projects), though there is no clear renewable heating target available.

To accomplish these goals, France has relied primarily on financial incentives and the use of regulations. There have been thermal regulations for new residential sector buildings since 1974 and, more recently, the White Energy Saving Certificate Scheme requires utilities to meet demand-side management targets in the residential and tertiary sector in proportion to the sales achieved in those sectors. The most important programme on the national level is a tax credit (*crédit d'impôt*) for the installation of renewable heating technologies. There are a number of national programmes to promote renewable heating installations and many regions also provide direct grants, mostly for solar thermal technologies.

Results

France has a moderate installed capacity of solar heating per capita and a relatively moderate rate of increase. The air-source heat pumps market is growing moderately in France while the ground-source heat pump market is growing more significantly. France has moderate biomass pellet burner population.

Significant programmes of interest

- Tax credit/crédit d'impôt.

6.5 Germany

Germany is one of the most advanced countries in renewable energy usage. Renewable heat is no exception; although biomass heating is the largest contributor to the renewable heating market in Germany, the solar thermal and geothermal heat pump markets have experienced rapid growth due to well-organized subsidies, regulations and general public awareness.

Context

Germany is a federation composed of 16 states (*Länder*). Energy policy is developed at the federal level, but the states are responsible for implementation.[30] Germany produces small amounts of oil and natural gas, hydroelectricity, nuclear power (powered by imported uranium) and coal, but relies extensively on imported oil and gas. Conventional energy prices are presented in Exhibit 5. The country has relatively low access to solar resources, moderate access to biomass resources and good access to low-temperature geothermal resources. The climate is temperate, with mild winters and cool summers.

Technology market status

Good quality solar thermal systems for residential applications are readily available at a cost of approximately €3000–5000 for hot water only and €8000–12,000 for a combined water and space heating system, including installation and taxes.[31] By 2007, 105,000 pellet appliances had been installed.[32] The pellet heating market increased dramatically between 2004 and 2007, however, the market slowed in 2008 due to volatility of pellet prices and supply. Germany is currently a leader in the European heat pump market[33] with 52,000 new installations in 2007.[34] Subsidies are shifting the market away from air-source heat pumps to ground-source heat pumps.

Programmes

The German government supports the liberalization of European energy markets, increased energy efficiency and diversification in energy production, with one priority being the use of renewable energies. Germany has adopted the EU renewable energy targets.

To accomplish these goals, Germany has relied primarily on financial incentives. However, recently this has been augmented by the use of regulations. The most important programme on the national level is the Market Incentive Programme (MAP). Since its start in 1999, it has focused on the promotion of renewable heat production (solar-thermal and biomass). Since 2009, MAP has been combined with the obligatory use of renewable energies in new residences (*EEWärmeG*).

Results

Germany has one of the highest installed capacities of solar heating per capita and the rate of increase continues to be one of the highest. A large market for biomass, especially wood pellets, has developed and is growing constantly. The installation of efficient heat pumps has only been promoted by MAP since 2008. Consequently the installed capacity and market growth are lower when compared to the solar and biomass markets.

Significant programmes of interest

- MAP – Market Incentive Programme (1999–present) biomass and solar; (2008) heat pumps.
- Marburg, Germany Ordinance.

6.6 Ireland

Prior to 1998, no renewable heating programmes existed in Ireland. New incentive programmes, aimed, in particular, at residential customers, have initiated significant growth in this sector. Unlike several other European countries, Ireland has set specific renewable heating targets.

Context

Ireland is a republic and parliamentary democracy divided into 26 counties. Energy legislation is designed and accepted at the federal level. The national energy agency, Sustainable Energy Ireland (SEI), advises the federal government on policies and measures on sustainable energy and implements the programmes developed by the federal government. Ireland produces small amounts of solid fuels (peat) and gas. (Peat is a significant residential heating source and is less expensive than both heating oil and gas.) No nuclear power is produced. Foreign energy dependency is high and most oil and gas is imported from the UK or from Norway.[35] Conventional energy prices are presented in Exhibit 5. Ireland has low access to solar resources, moderate access to biomass resources, and good access to low-temperature geothermal resources. Ireland is temperate with an oceanic climate that has few extremes and lots of precipitation.

Technology market status

The thermal solar system market in Ireland lagged behind that in many other countries until very recently: only 3500m^2 were installed during 2005. In 2007, however, 10,000m^2 of flat collectors and 5000m^2 of vacuum collectors brought the total solar thermal capacity in Ireland to 30,800m^2 (21,550kWth).[36] Most systems are imported because there is almost no domestic production. The market for pellet heating in Ireland started in 2006. Although peat has been used traditionally, wood was never a popular heating fuel due to lack of availability. Most heat pumps installed are ground-source, although the market is still at an early stage.

Programmes

Due to the high energy import dependency, the main aim of the Irish energy policy is to ensure the security of supply. While respecting the EU objective of 20 per cent renewable energy by 2020, Ireland has specific targets for heating from renewable energy sources: 5 per cent market penetration of renewables in the heat market by 2010 and 12 per cent by 2020.

Specific mechanisms used to further these policy targets are mostly financial incentives and awareness campaigns through SEI. A regulation has recently been implemented regarding the use of renewable energy resources in new buildings. The technical guidance for this regulation states that a 'reasonable minimum level of energy provision from renewable energy technologies in order to satisfy Regulation L2 (b)' can be '10kWh/m^2/annum contributing to energy consumption for domestic hot water heating, space heating or cooling'.[37] No programmes for renewable energy heating were in existence before 1998, with most programmes beginning around 2005.

Results

It is too early to give results of the programmes implemented for heating from renewable sources. Current renewable energy heating capacities are low compared to other European countries. However, significant market growth has been recorded for renewable heating technologies since the implementation of the newer programmes.

Significant programmes of interest

- Buildings regulation part L amendment (2008–present).
- GreenerHome (2006–present).

6.7 Italy

Despite relatively good policies in Italy for renewable energy, including numerous incentive schemes, the renewable energy share of gross final energy consumption has remained relatively stable.

Context

Italy is a republic composed of 20 regions. The government has federal-level energy policies, but also allows the regions to develop their own programmes with regard to energy, including renewables. The country produces some gas and oil, but relies extensively on imported oil and gas.[38] Conventional energy costs are presented in Exhibit 5. The country has good access to solar resources, moderate access to biomass resources and good access to low-temperature geothermal resources. Italy has a Mediterranean climate that is more temperate in the north.

Technology market status

The mild climate in Italy allows the installation of less-expensive solar systems (€3500–4000 including taxes and installation).[39] The pellet stove market is robust even though there are no specific programmes promoting this technology. The growth appears to be due to high conventional energy prices and easy availability of stoves and boilers. Due to the mild climate, most heat pumps are air-sourced.

Programmes

The current energy policy in Italy covers issues such as: emphasis on energy diversification while respecting environmental concerns: electricity market liberalization; increasing energy security; and reducing energy prices.[40] In addition, recent changes to the energy policy programme for 2009 include efficiency improvements in energy generation; enhanced application of cogeneration, efficiency and energy saving technologies; and renewable energies.[41] National targets follow European policies, for the most part (with a target of a 17 per cent share of renewable energy in final energy consumption), but there are no specific renewable heating targets in place.

   To accomplish these goals, Italy uses financial and tax credit incentives and is working to improve the regulatory framework, mainly in terms of procedure simplification. There are national market-based incentives (such as white certificates)[42] and fiscal incentives (55 per cent tax deduction)[43] to promote renewable heating technologies as well as various regional subsidy programmes such as the Kyoto Revolving Fund[44] and the Regional Operational Programme under the EU Policy for Local Government. Regulations requiring the use of renewable heat in new and renovated buildings, both public and private, were established in 2006.[45]

Results

Italy has a moderate installed capacity of solar heating per capita and a good rate of market growth in the last few years.[46] The market for air-source heat pumps grew significantly in Italy, but few ground source heat pumps are currently installed. Despite having only moderate biomass resources, Italy has a significant number of installed pellet burners. The driver for biomass heating appears to be the economic advantage when compared to the price of conventional energy sources.

Significant programmes of interest

- Tuscany biomass regional subsidy programme.
- Italian Tax Credit and VAT reduction.

6.8 Japan

Japan has limited domestic fossil fuel resources and as a result has made great gains in energy efficiency. The country does, however, have significant geothermal resources and technical potential. It had a period of great solar thermal market growth that has since stagnated. Heat pumps, on the other hand, are growing in popularity, due to their appropriateness for the climate and recent promotional action from the government.

Context

Japan is a constitutional monarchy divided into eight distinct regions. Energy policy is controlled both by the federal and the municipal level. Japan has limited energy resources and is highly reliant on coal and oil imports.[47] Kerosene is an important heating fuel because it is easy to use in both portable and installed space heaters. (Japanese homes do not usually have central heating.) Conventional energy prices are presented in Exhibit 5. The country has moderate access to solar resources, low access to biomass resources and good access to low temperature geothermal resources. The climate is generally temperate, but varies from long, cold winters and cool summers in the north to a subtropical climate with a rainy season on the southern islands.

Technology market status

Solar thermal heat in Japan has a long history, however, it stagnated in the late 1990s due to lower petroleum prices and a strong Japanese currency.[48] Information about residential-scale biomass heating in Japan is difficult to locate. Japan has traditionally used direct geothermal heat for a variety of applications. Geothermal heat has only recently been promoted under renewable energy programmes and heat pump use is now increasing.

Programmes

As a country very reliant on energy imports, Japan has survived two oil shocks by increasing energy efficiency and decreasing energy consumption per GDP. The main objectives of Japanese energy policy are the development of a state-of-the-art energy supply and demand structure, strengthening resource diplomacy and energy and environmental cooperation, enhancing emergency response measures and enhancing technology R&D.[49] The New Energy Law (1997–present) policy promoted oil-alternative resources including renewable technologies, hydrogen technologies, combined heat and power (CHP) generation and waste power. Revisions in 2002 included biomass and cold energy (snow/ice). Geothermal energy is not currently included. Under the New Energy Law, subsidies are provided to homeowners and support is offered to energy-related businesses. This has been mostly focused on solar businesses, but more recently, a specific incentive for biomass has been implemented. Support for heat pumps is limited, to some energy efficiency subsidies.

At the municipal level, the Tokyo metropolitan government Initiative has also set targets: 20 per cent renewable energy, with a focus on renewable heating, by 2020.

Results

The growth of the solar thermal heat market has been slowing since the 1990s, after 30 years of strong solar thermal development, due to a decrease in the price of conventional fuel and heating technologies. Despite a return to higher energy prices, the solar hot water market has continued in a downward trend. The biomass heating market is limited due to scarce local resources. Despite the amount of geothermal activity in Japan, heat pump development has been limited due to its exclusion from the New Energy Law.

Significant programmes of interest

- Law on Special Measures for the Utilization of New Energy (photovoltaics, solar thermal, wind, waste power, biomass) (2002).

6.9 Netherlands

The Netherlands has established overarching national environmental objectives that have promoted renewable energy and altered the behaviour of business and industry. This includes a significant increase in the amount of biomass used for cogeneration of heat and power plants (CHP). Despite this, the country offers somewhat limited support to renewable heat for residential buildings.

Technology market status

The solar water heating market was unstable in 2005 and 2006 with installations of 20,248m² and 16,685m² respectively. In 2007, another 19,900m² was installed, leading to a total of 338,341m² (236,839kWth) in operation in 2007.[50] The average price of a solar system for water heating is €1800–2700, including taxes and installation.[51] The biomass market is not developed in the Netherlands. The heat pump market is predominantly composed of air-source heat pumps. Approximately 29,000 heat pumps had been installed in households as of 2007.[52]

Context

The kingdom of the Netherlands is a constitutional monarchy divided into 12 provinces. Energy policies and targets are developed and controlled at the federal level. The Netherlands produces and uses large amounts of natural gas, but has a significant import dependency on oil. Smaller amounts of oil and nuclear power are produced.[53] Conventional energy prices are presented in Exhibit 5. The country has relatively low access to solar resources, low access to biomass resources and good access to low-temperature geothermal resources. It has a temperate, somewhat maritime climate, with cool summers and warm winters.

Policy

The Dutch strategy includes the optimized use of domestic energy sources, namely natural gas, the security of access to foreign gas resources and the intensification of international cooperation in energy questions.[54] The Netherlands has adopted the EU renewable energy targets. In addition, the Netherlands has proposed to increase renewable heat production to 3 per cent of total energy supply, mainly through heat pumps and solar hot water systems.[55]

The Netherlands offers financial incentives and has encouraged the inclusion of solar thermal heating or heat pumps in new buildings through programmes such as energy performance requirements. Biomass for heating is largely neglected because the forestry industry is limited and mainly based on imported wood. The Dutch National Action Plan[56] mentions small-scale wood combustion only briefly and states that its (low) level should be kept by increasing the efficiency of these applications. Furthermore, small-scale wood heating is not included in the national support programmes.

Results

The Netherlands has limited installed capacity and market growth for solar thermal heating. Residential biomass heating has had negligible market penetration. Heat pumps have also had limited implementation, however, the new building regulations are expected to encourage the use of geothermal and air-source heat pumps.

Significant programmes of interest

- Energy Performance Requirements For New Construction.
- Solar Domestic Hot Water System Agreement.

6.10 Norway

Norway is a country rich with renewable and non-renewable natural resources. Despite large exports of fossil fuels, Norway's main source of energy is hydroelectricity. Although conservation efforts have not been as vigorous as in other Nordic countries, Norway has used carbon taxes and incentives to encourage renewable energies. In particular, biomass and heat pump technologies have been encouraged.

Context

Norway is a constitutional monarchy with a parliamentary system of government and is composed of 19 administrative regions (*fylker*). The *fylker* have limited authority, and most energy policy is administered at the national level. The country is mostly reliant on hydro power for electricity production as well as small amounts of petroleum products. Norway exports great quantities of petroleum and natural gas.[57] Conventional energy prices are presented in Exhibit 5. Prices of conventional energy sources are low. The country has relatively low access to solar, and good access to traditional wood biomass resources and low-temperature geothermal resources. Although the northern mainland of Norway has a sub-Arctic climate, the southern and western parts are warmer. Inland, winters are colder with less precipitation. Daylight hours vary greatly from winter to summer.

Technology market status

Norway has not aggressively promoted solar thermal heating. Biomass stoves are the second largest source of household heat, after electricity. Only 18 per cent of these stoves are energy efficient, and very few are pellet stoves.[58] Heat pumps are not well-suited to the coldest areas of Norway, however there has been a tenfold increase in sales of brine to water heat pumps from 1996–2006, and a 100-fold increase in sales of air to air heat pumps during the same period.[59]

Programmes

Energy conservation efforts have been less aggressive than those in other Scandinavian countries due to the abundance of hydroelectric power. The government's renewable energy development objectives are significant growth in renewable energy production and increased energy efficiency with the short-term targets of: significantly increasing renewable district heat production to a minimum of 4TWh annually; increasing wind energy production to 3TWh by the end of 2010 as stated in the 1998–1999 federal white paper on energy policy;[60] and reducing market barriers.[61]

Norway's main programme is *Energi 21*, established in 2007 to provide a platform for supply-side security in the energy sector by promoting and coordinating a commitment to research, development, demonstration and commercialization of new technologies. Incentives include project funding as well as grants for the installation of specific technologies. In addition, a regulation was imposed in 1999, mandating a connection to district heating where available.

Results

Norway has a low installed capacity of solar heating per capita and a low rate of increase. There are a relatively high number of air source heat pumps in Norway and a moderate number of ground source heat pumps. Traditional wood combustion is widespread due to the ample biomass resources available; although almost two-thirds of homes have biomass heating of some sort, it is rarely the primary heat source.[62] Modern biomass, including pellet technologies, is more rare.

Significant programmes of interest

- Household Subsidy Programme for electricity saving heating technologies.

6.11 Spain

Spain lacks conventional energy resources, with the exception of coal. Although the country has had considerable economic growth and construction, energy consumption and intensity have been decreasing for the last four years due to conservation and efficiency policies.[63] The government has been using policy to increase the use of electricity from renewable sources. However, renewable energy exploitation for heating and cooling is limited. Regional incentives and regulations have been more successful, in particular those in Andalusia, Barcelona, Catalonia and the Canary Islands.

Context

Spain is a constitutional monarchy composed of 17 regions. The energy sector is regulated at national level. Each state administrates procedures and provisions related to the environment and regional authorities are responsible for some renewable energy planning.[64] The country is heavily reliant on natural gas and nuclear energy for electricity production and has a significant renewable energy contribution. Spain also relies extensively on imported oil and gas. Exhibit 5 presents conventional energy prices. The country has relatively good access to solar resources, low access to biomass resources and good access to low-temperature geothermal resources. Spain has a Mediterranean climate in the south, with colder winters and warmer summers in the inland area of the peninsula.

Technology market status

In spite of Spain's excellent solar resources, the solar thermal market grew very slowly from 1990–2005, with the exception of certain regions with municipal programmes.[65] In 2006, regulations began that required solar thermal in many new and renovated buildings. Energy efficient biomass heating in Spain is only starting to develop. In 2008, approximately 1000 pellet boilers were installed and the annual pellet consumption was less than 10,000 tonnes.[66] Air-source heat pumps are more popular in Spain than elsewhere in Europe,[67] however, ground-source heat pump market is small.

Programmes

The government supports opening energy markets in the European framework, increased energy use efficiency, investment in national gas and electricity infrastructure, and enhanced use of renewable energy sources and cogeneration. National targets include the reduction of energy consumption by 8.5 per cent and energy imports by 20 per cent by 2012.[68] In addition, Spain has set the specific objective that 12 per cent of all primary energy sources be renewable by 2010.[69]

To accomplish this goal, Spain uses a regulatory instrument, namely, the federal Technical Building Code CTE, which requires the installation of solar thermal technologies in new buildings and buildings under renovation. Barcelona and Pamplona have also implemented successful solar ordinances. In addition, financial incentives have been used to promote the installation of renewable technologies (solar thermal and biomass) in the residential sector.

Results

Spain has a moderate installed capacity of solar heating per capita and a relatively moderate rate of increase. Air source heat pumps are popular in Spain due to the country's suitable warm climate, but the country has a relatively low number of ground source heat pumps installed. The number of pellet burners is unknown.

Significant programmes of interest

- Barcelona's Solar Thermal Ordinance (2000).
- Technical Building Code through PER (2005–present) – for solar thermal.

6.12 UK

The UK has made some progress in terms of renewable energy technologies and, with respect to residential heating, is primarily trying to address energy poverty problems. The country has rather limited renewable resources available. However, a policy is in development that could greatly increase the renewable heat market.

Context

The UK is a constitutional monarchy with a parliamentary government. It has devolved national administrations including Northern Ireland, Scotland and Wales, which have their own government or Executive. Energy policy is not devolved to Wales, so renewables policies for Wales and England are defined by the UK national government. Scotland aligns its policy with England and Wales, whereas Northern Ireland tends to align its policies with Ireland. The UK traditionally produces coal, natural gas and petroleum. Coal has been the most important fuel for electricity production, however production has fallen significantly since 1980 (due to depletion of easily-accessed reserves) and imports have increased. Natural gas has increased in importance, and there is a significant gas network offering an easily accessible and relatively cheap fuel for heating. Nuclear and oil are less important fuel sources. Conventional fuel prices are presented in Exhibit 5. The country has relatively low access to solar resources, low access to biomass resources and good access to low-temperature geothermal resources. Climate is temperate with year-round rainfall, but temperatures seldom fall below -10°C.

Technology market status

The thermal solar systems and energy-efficient biomass markets are not well-developed in the UK. There was a sharp increase in the small heat pump market in 2007, mainly due to higher energy prices.[70]

Programmes

The main components of UK energy policy are: to work towards reducing CO_2 emissions by 60 per cent, with verifiable results by 2020; to maintain reliable energy supplies; to promote competitive markets in the UK and abroad to encourage sustainable economic growth and to improve productivity; and to ensure that every home is adequately and affordably heated.[71] Quotas for electricity (from licensed suppliers) from renewable sources are set annually, starting at 3 per cent in England, Wales and Scotland, rising to 15.4 per cent in 2015–2016.[72] In July 2009, the Renewable Energy Strategy document set a target of producing 12 per cent of all energy used to provide heat with renewable energy by 2020.[73]

The main mechanisms for implementing these policies are financial incentives and regulations. The 2003 Clear Skies Initiative offered support for biomass and solar heating. This was eventually replaced by the Low Carbon Buildings Programme, which encourages microgeneration technologies including bioenergy, heat pumps and solar heating. This programme was set to end during the spring of 2010. The UK Renewable Heat Incentive is a new programme scheduled to begin in April 2011, which will offer incentives for renewable heat generated at household, community and industrial levels, including a feed-in tariff for injecting biomethane into the gas grid.[74] It seems that there will be a gap between the Low Carbon Buildings Programme and the Renewable Incentive Heat Programme of approximately one year.

One guidance measure is the Microgeneration Certification Scheme, which applies to manufacturers and installers of renewable energy equipment. This scheme has set up voluntary standards for the three types of renewable heat technologies.

Results

Solar thermal has increased in the last few years as a result of the Clear Skies Initiative, however, the market has been limited by various factors including lack of installers. Energy efficient biomass heating is still quite limited. The heat pump market is growing slowly. High expectations have been set for the Renewable Heat Incentive in terms of increase in national renewable heating capacity.

Significant programmes of interest

- Microgeneration Certification Scheme.
- Solar for London.
- Clear Skies Programme (2003–2006).

6.13 US

Certain regions of the US are more advanced in renewable energy use, depending on their energy, climate and political situations. California and Hawaii, where the solar resource is abundant, are the strongest in solar applications. The northeast, however, is the showing most interest in biomass for heating to offset expensive heating oil. Market growth for air-source heat pumps is mostly focused in the midwest, due to promotion by utility companies.

Context

The US is a federal constitutional republic made up of 50 states and a federal district. Energy programmes are organized at both the federal and state level, however, the state governments have their own taxes and significant authority over energy matters. The US produces coal, natural gas, oil, hydroelectric power and nuclear power. Nuclear power has been less utilized in the US than in many other developed countries. Prices of conventional energy sources are presented in Exhibit 5. The country has good access to solar resources, good access to biomass resources and good access to low-temperature geothermal resources. The climate varies from a humid temperate climate with cool winters and warm summers to subtropical in the south.

Technology market status

Interest in thermal solar systems varies greatly by region, solar resource, climate and political environment. California and Hawaii both have well-developed programmes. Wood heating is well-established in some regions of the US. The EPA Certified Stove regulations were primarily aimed at limiting emissions from woodstoves but have had the additional effect of increasing efficiency. The pellet stove market is limited. Heat pump use varies by climate and programme availability. The ground source heat pump market was strong until 2000 but has since slowed.[75]

Programmes

The US government supports energy security, nuclear security, scientific discovery and innovation, environmental responsibility and excellence in management. The 2009-issued New Energy for America Plan includes: reducing fossil fuel imports from the Middle East and Venezuela within ten years; ensuring 10 per cent of electricity comes from renewable sources by 2012, and 25 per cent by 2025; reducing greenhouse gas emissions by 80 per cent by 2050 through an economy-wide cap-and-trade programme; and making the US a leader on climate change.[76] Individual states have also implemented aggressive policies related to climate change, particularly Hawaii and California.

The US has relied primarily on financial incentives in terms of tax credits at the federal and state level. Other rebate programmes are available at the regional level through utility companies that provide loans and other subsidies. Also, national labelling programmes have been instrumental in promoting energy efficient biomass technology. Regulations have been developed particularly for solar thermal installations in Hawaii, and installer certification programmes exist in Hawaii and California.

Results

The US has had great success in specific states with different renewable energy technologies. Solar is supported by a variety of subsidies at state and regional levels in Hawaii and California, and has resulted in significant market growth of solar thermal in those areas. Woody biomass is a traditional fuel in the US, and voluntary labelling programmes through the Environmental Protection Agency have increased the presence of energy efficient biomass heating technologies in the market. The geothermal heat pump market shows significant growth.

Specific programmes of interest

- California Global Warming Solutions Act of 2006 Policy (2006–present).
- Building Technologies Programmes – tax credits and standards (2005–present).
- Indiana utility-level geothermal and air-source heat pump incentives.
- California Solar Contractor Licensing.
- EPA certification and guidelines for heat pumps and biomass stoves.
- Hawaii Solar Water Heating for New Construction.
- California Solar Rights Act.
- South Carolina Solar Heating and Cooling Personal and Tax Credit.

6.14 China

China has demonstrated clear leadership in total capacity of solar thermal glazed flat plate and evacuated tube collector installations (79,898.0MWth by the end of 2007), more than ten times the installations of Turkey, the country with the second most MWth of solar thermal capacity (7105.0MWth). This is partly due to the much larger population of China compared to any other country: a per capita analysis of solar thermal installations reveals that China is the tenth-ranked country, at 60.1kWth installed per 1000 inhabitants in 2007. However, installed per capita capacity is currently increasing rapidly.[77] This solar boom in China developed in the absence of any government incentives. This is probably because of several major advantages of solar in China:

- High solar resources: 70 per cent of the country has solar resources in excess of 500kJ/cm^2/year.
- Inexpensive installation, at around $0.25/watt (compared to $2/watt in most countries).[78]
- The large number of manufacturers in China (in 2003, 75 per cent of solar thermal collectors produced worldwide were made in China).[79]

- Higher competition for conventional energy sources and increased electricity and heat demand from the rapidly growing economy.
- Opportunity to mitigate some of the greenhouse gas emissions resulting from the increasing demand for energy.

In terms of national policy, the government in China has recently become active in the promotion of renewable energy for electricity and heating. The government has set a target of 15 per cent of energy output from renewable sources by 2020. Solar thermal is the second largest renewable energy source with a contribution of approximately 5MToe (209PJ); it is predicted this will rise to 20MToe (837PJ) by 2020.[80]

The main barriers for the further development of the Chinese solar water heater market are:[81]

- The low-level technology and limited innovation in the solar water heater system manufacturing market.
- The limited technical service system: manufacturers outnumber technical service agencies 4:1.
- The lack of national production quality control: the eight national and industrial standards that have been established are not enforced.
- Poor design: new home designers often do not take into account the design of solar water heater equipment and the roof-top installation area, and therefore the equipment cannot fit well on newly constructed homes.
- Lack of awareness: until very recently, public awareness about energy conservation and environmental protection has been limited, and therefore selection of water heaters (solar or otherwise) was only based on price and product performance.

Notes

1   *Energieflussbild Österreich 2005*, Austrian Energy Agency, www.energyagency.at, accessed January 2009.
2   *Natural Gas*, OVGW, www.ovgw.at/en/themen/index_html?uid:int=175, accessed June 2009.
3   *Austria Energy Mix Fact Sheet*, European Commission, 2007, www.energy.eu/renewables/factsheets/2008_res_sheet_austria_en.pdf, accessed June 2009.
4   Faninger, G., *Der Solarmarkt in Österreich 2006*, BMVIT, 2007.
5   *Basisdaten Bioenergie Österreich 2006*, Energyagency Austria, 2007.
6   Pellets @las website, www.pelletsatlas.info, accessed January 2009.
7   Faninger, G., *Aktueller Stand der Wärmepumpentechnik in Österreich*, Universität Klagenfurt, 2007.
8   'Heat pumps barometer', *SYSTÈMES SOLAIRES: Le journal des énergies renouvelables*, no 193, 2009. Available at www.eurobserv-er.org/pdf/baro193.pdf.
9   *Regierungsprogramm 2007–2010,* Bundeskanzleramt Österreich, 2007.

10 *Austria Renewable Energy Fact Sheet,* European Commission, 2008. Available at
    www.energy.eu/renewables/factsheets/2008_res_sheet_austria_en.pdf.

11 *2007 Energy Balance for Canada*, IEA, www.iea.org/Textbase/stats/balancetable.
    asp?COUNTRY_CODE=CA, accessed February 2009.

12 Marbek in-house data.

13 Bradley, D., *Canada Report on Bioenergy 2008*, Ontario: Climate Change Solutions,
    2008. Available at www.bioenergytrade.org/downloads/
    canadacountryreportjun2008.pdf.

14 Personal communication with Al Clark, programme manager for the Renewable
    Heat Programme of Natural Resources Canada, May 2010.

15 *Canada Lists Emissions Target under the Copenhagen Accord*, Environment
    Canada. www.ec.gc.ca/default.asp?lang=En&n=714D9AAE-1&news=EAF552A3-
    D287-4AC0-ACB8-A6FEA697ACD6, accessed May 2010.

16 ecoEnergy for Renewable Heat and ecoEnergy Retrofit programmes are too new to
    review. The completed REDI programme has been evaluated.

17 Hasselager, A., *Danish Energy Policy: Focus on Heat and Power Production and
    Renewable Energy*, Danish Energy Authority, 2008. Available at
    http://74.125.95.132/search?q=cache:SedC7odPfhoJ:www.ambsofia.um.dk/NR/
    rdonlyres/4B2B02C9-4E71-445C-817B-FBA50D8BD8C1/0/
    PresentationAndersSofiaworkshop1.ppt+energypolicy+denmark+region&hl=en&ct
    =clnk&cd=3&gl=ca.

18 *Denmark – Energy Mix Fact Sheet*, European Commission, 2007. Available at
    http://ec.europa.eu/energy/energy_policy/doc/factsheets/mix/mix_dk_en.pdf.

19 Stenkjaer, N., *Solar Thermal Collectors,* Nordic Folkecenter for Renewable Energy,
    www.folkecenter.net/gb/rd/solar-energy/solar_collectors, accessed February 2009.

20 *Large and Small Scale District Heating Plants: Individual Heat Installations (Houses,
    Farms),* Danish Energy Agency, www.ens.dk/en-US/supply/Heat/district_heating_
    plants/Sider/Forside.aspx, accessed February 2009.

21 *Heat Pumps,* Danish Energy Agency, 2006, www.ens.dk/sw20505.asp, accessed
    February 2009.

22 'Heat pumps barometer', *SYSTÈMES SOLAIRES: Le journal des énergies
    renouvelables*, no 193, 2009. Available at www.eurobserv-er.org/pdf/baro193.pdf.

23 *Energy Policy Statement 2008,* Danish Energy Agency, www.ens.dk/graphics/
    Publikationer/Energipolitik_UK/energipolitisk_redegorelse_2008_eng/html/kap01.
    htm, accessed February 2009.

24 IEA-RETD, *Renewables for Heating and Cooling*, Paris: OECD/IEA, 2007. Available
    at www.iea.org/publications/free_new_Desc.asp?PUBS_ID=1975.

25 EDG website, www.edg-shop.de/artikelueber.php?wgruppeid=921&wgruppe_offe
    n=921&sid=9a8b20c32f7de07cf5ae3260e8d25283, accessed February 2009.

26 *Enquête sur les ventes d'appareils domestiques de chauffage au bois en 2006*,
    ADEME, 2007.

27 Douard, F., *Challenges in the Expanding French Pellet Market*, ITEBE, 2007, www.
    bioenergynews.org/images/pdf/Challenges_expanding_French_pellet_market_
    Wels_2007.pdf, accessed February 2009.

28 *Outlook 2008: European Heat Pump Statistics/Summary*, Brussels: European Heat
    Pump Association, 2008.

29 'Heat pumps barometer', *SYSTÈMES SOLAIRES: Le journal des énergies renouvelables*, no 193, 2009. Available at www.eurobserv-er.org/pdf/baro193.pdf.

30 *Energy Policies of IEA Countries-Germany 2002 Review*, Paris: IEA, 2002. Available at www.iea.org/textbase/nppdf/free/2000/germany2002.pdf.

31 SolarOne Deutschland website, www.solarone.de, assessed January 2009.

32 Deutscher Energie-Pellet-Verband website, www.depv.de, accessed January 2009.

33 IEA-RETD, *Renewables for Heating and Cooling,* Paris: OECD/IEA, 2007. Available at www.iea.org/publications/free_new_Desc.asp?PUBS_ID=1975.

34 Walker-Hertkorn, S., *Marktentwicklung Wärmepumpen*, Bundesverband WärmePumpe e.V. 2008.

35 *Ireland – Energy Mix Fact Sheet*, European Commission, 2007. Available at http://ec.europa.eu/energy/energy_policy/doc/factsheets/mix/mix_ie_en.pdf.

36 *Solar Thermal Markets in Europe: Trends and Market Statistics 2008*, Brussels: European Solar Thermal Industry Federation, 2008. Available at www.estif.org/fileadmin/estif/content/market_data/downloads/2008%20Solar_Thermal_Markets_in_Europe_2008.pdf.

37 Environment, Heritage and Local Government website, www.environ.ie/en/TGD, accessed June 2009.

38 *Italy – Energy Mix Fact Sheet*, European Commission, 2007. Available at http://ec.europa.eu/energy/energy_policy/doc/factsheets/mix/mix_it_en.pdf

39 Consultation with retailer.

40 Capra, M., *The Italian Policy with Reference to the European Community Trends*, Sustainable Fossil Fuels for Future Energy, 2009, www.co2club.it/agenda/Presentations/M.Capra_S4FE_070709.pdf, accessed August 2009.

41 Kovalyova, S., *Update 1 – Italy to Present Energy Policy in Spring*, Reuters, September 20 2008, http://uk.reuters.com/article/idUKLK19708420080920, accessed July 2009.

42 *Law Decrees on 20 July 2004.*

43 Established by the 2007 Financial Law.

44 Established by the 2007 Financial Law (commas 1110–1115).

45 IEA-RETD, *Renewables for Heating and Cooling*, Paris: OECD/IEA, 2007. Available at www.iea.org/publications/free_new_Desc.asp?PUBS_ID=1975.

46 *Solar Thermal Markets in Europe: Trends and Market Statistics 2008*, Brussels: European Solar Thermal Industry Federation, 2008. Available at www.estif.org/fileadmin/estif/content/market_data/downloads/2008%20Solar_Thermal_Markets_in_Europe_2008.pdf.

47 *2006 Energy Balance for Japan*, IEA, www.iea.org/Textbase/stats/balancetable.asp?COUNTRY_CODE=JP, accessed February 2009.

48 IEA-RETD, *Renewables for Heating and Cooling*, Paris: OECD/IEA, 2007. Available at www.iea.org/publications/free_new_Desc.asp?PUBS_ID=1975.

49 Hombu, K., *Japan's Energy Policy and Japan-India Energy Cooperation*, DDG for Energy and Environment, METI, 2006, www.nedo.go.jp/kokusai/kouhou/181206/session01/1-1.pdf, accessed February 2009.

50 *Solar Thermal Markets in Europe: Trends and Market Statistics 2008*, Brussels: European Solar Thermal Industry Federation, 2008. Available at www.estif.org/fileadmin/estif/content/market_data/downloads/2008%20Solar_Thermal_Markets_in_Europe_2008.pdf.

51   *De anschaaf van en zonneboiler*, Holland Solar.

52   *Duurzame Energie in Nederland 2007*, Centraal Bureau voor de Statistiek, 2008.

53   *The Netherlands – Energy Mix Fact Sheet*, European Commission, 2007. Available at http://ec.europa.eu/energy/energy_policy/doc/factsheets/mix/mix_nl_en.pdf.

54   *Energierapport 2008*, Dutch Ministry of Economic Affairs, 2008.

55   *Part 5: The Dutch National Environmental Plans,* Renewable Energy Policy Project, 1999, www.repp.org/repp_pubs/articles/issuebr14/05Dutch.htm, accessed February 2009.

56   Dutch Ministry of Economic Affairs, Biomass Action Plan – The Netherlands (2003–2006), 2003.

57   *Energy Fact Sheet Norway*, RICS research, 2008, www.rics.org/NR/rdonlyres/7FDEB8E4-3834-4014-B4E2-F5BA6B9367F1/0/Norwayenergyfactsheet.pdf, accessed May 2009.

58   Reenaas, M. et al, *Life Cycle Assessment of Wood Based Heating in Norway*, Norwegian University of Science and Technology, 2006, www.dtu.dk/upload/subsites/norlca/proceedings%202006/reenaas%20et%20al.pdf, accessed February 2009.

59   Mdittomme, K. et al, *Ground-source Heat Pumps and Underground Thermal Energy Storage – Energy for the Future*, Geological Survey of Norway (NGU). Available at www.ngu.no/upload/Publikasjoner/Special%20publication/SP11_08_Midttomme.pdf.

60   *Stortingsmelding 29* (1998–1999).

61   *External factors – Norway: Norwegian Energy Policy*, www.hsaps.ife.no/Presentations_pdf/external%20factors%20Norway.pdf, accessed February 2009.

62   Reenaas, M. et al, *Life Cycle Assessment of Wood Based Heating in Norway*, Norwegian University of Science and Technology, 2006. Available at www.dtu.dk/upload/subsites/norlca/proceedings%202006/reenaas%20et%20al.pdf, accessed February 2009.

63   Marin, P., *La intensidad energetic cayo en 2008 por cuarto ejercicio consecutive gracias a las politicas de ahorro y eficiencia energetic*, Gobierno de Espana: Ministrio de Industria, Turismo y Comercio, 2009, www.mityc.es/es-ES/GabinetePrensa/NotasPrensa/Paginas/npcomparecenciamarincongreso.aspx, accessed August 2009.

64   *Renewable Energy Policy Review: Spain*, Brussels: European Renewable Energy Council, 2004. Available at www.erec.org/fileadmin/erec_docs/Projcet_Documents/RES_in_EU_and_CC/Spain.pdf.

65   Amblar, P. P., *Promotion of Heating and Cooling from Renewable Energy Sources in Mediterranean Countries*, ASIT, 2007.

66   Puente, S., *Pellets Market in Spain,* Pro Pellets Spain, Wels, 2009.

67   Dewdney, P., *Heat Pump News Issue: 1 Autumn 2002*, Heat Pump Association. Available at www.heatpumps.org.uk/PdfFiles/HeatPumpNewsNo.1.pdf.

68   *La Energia en Espana*, Ministerio de Industria, Turismo y Comercio, 2007.

69   IEA-RETD, *Renewables for Heating and Cooling*, Paris: OECD/IEA, 2007. Available at www.iea.org/publications/free_new_Desc.asp?PUBS_ID=1975.

70   *Heat Pump Market Growing Fast,* BSRI, 2008, www.bsria.co.uk/news/heatpump08, accessed February 2009.

71 *Energy White Paper: Meeting the Energy Challenge*, BERR, 2007, www.berr.gov.
uk/whatwedo/energy/whitepaper/page39534.html, accessed February 2009.

72 Renewable Energy Association website, www.r-e-a.net, accessed February 2009.

73 Connor, P. and Xie, L., *Current State of Heating and Cooling Markets in United
Kingdom*, Exeter: Intelligent Energy Europe, 2009. Available at www.res-h-policy.
eu/downloads.htm.

74 *Renewable Heat Incentive (RHI)*, BERR, 2009, www.berr.gov.uk/energy/sources/
renewables/policy/renewableheatincentive/page50364.html, accessed August 2009.

75 Hughes, P. J., *Geothermal (Ground-Source) Heat Pumps: Market Status, Barriers to
Adoption, and Actions to Overcome Barriers*, Oak Ridge: Oak Ridge National
Laboratory, 2008. Available at www.eere.energy.gov/geothermal/pdfs/ornl_ghp_
study.pdf.

76 *Energy and the Environment,* The White House, www.whitehouse.gov/agenda/
energy_and_environment, accessed February 2009.

77 Weiss, W. et al, *Solar Heat Worldwide: Markets and Contribution to the Energy
Supply 2007*, 2009 edition, Gleisdorf: IEA Solar Heating and Cooling Programme,
2009. Available at www.iea-shc.org/publications/statistics/IEA-SHC_Solar_Heat_
Worldwide-2009.pdf.

78 Bai, C., *Solar Energy Demand in China: Government Targets 15% Renewables by
2020*, ReneSola, www.renesola.com/media/ChinaSolarDemand.pdf, accessed
June 2009.

79,80 IEA-RETD, *Renewables for Heating and Cooling*, Paris: OECD/IEA, 2007. Available
at www.iea.org/publications/free_new_Desc.asp?PUBS_ID=1975.

81 Sizheng, H., *Solar Thermal Utilization in Testing,* OPET-CHINA Zhejiang Energy
Research Institute, www.fz-juelich.de/ptj/projekte/datapool/page/1134/Solar_
China_e.pdf, accessed June 2009.

7

Programme Selection and Case Studies

7.1 Programme selection

Selected Programmes	Single, detached house	Multi-unit dwellings	New construction	Retrofit of existing buildings
1. Energy Action Plan of Upper Austria	✓	✓	✓	✓
2. Climate Alliance of European Cities	✓	✓	✓	✓
3. Market Incentive Programme (Germany)	✓	✓	✓	✓
4. Regional programmes in Austria (three in total)	✓	✓	✓	✓
5. Household Subsidy Programme (Norway)	✓		✓	✓
6. Direct tax credit (France)	✓	✓	✓	✓
7. Barcelona Solar Thermal Ordinance (Spain)	✓	✓	✓	✓
8. Regional promotion programmes in Austria (two in total)	✓	✓	✓	✓
9. Solar Keymark (Europe)	✓	✓	✓	✓

Exhibit 14 Sector types

Candidates	Uptake	Continuity	Uniqueness	Data Availability	Scale	Comments
Regional subsidy programmes in Austria, such as the Salzburg, Steiermark and Vrarlberg programmes	✓	✓	✓	✓	✓	**Included**
Regional promotion programmes in Austria, such as *Spar mit Solar* and *Ja Zu Xolar*	✓	✓	✓	✓	✓	**Included**
Residential Solar Obligation (Austria)			x		✓	Excluded because the included Barcelona Solar Thermal Ordinance has been in existence for a longer time period
Klima : 'Wärmepumpe' , 'Holzwärme', 'Solarwärme' (Austria)			x	x		Excluded because the regional programmes in Austria were similar and more information was available
Energy Action Plan of Upper Austria	✓	✓	✓	✓		**Included**
Residential Earth Power Loan (Manitoba – Canada)	✓		x			Excluded because the included Enova programme is very similar
REDI Programme (Canada)		✓				Excluded since it is not specifically aimed at homeowners
Executive Order Solar Heating Obligation (Denmark)	x				✓	Excluded because programme has not met expectations
Danish Biomass Agreement		✓			✓	Excluded because district heating programmes are outside of scope
Direct Tax Credit (France)	✓	✓		✓	✓	**Included**
Marburg, Germany Ordinance	x	x	✓	x	x	Excluded because the ordinance is not in force yet due to legal constraints

Candidates	Uptake	Continuity	Uniqueness	Data Availability	Scale	Comments
Market Incentive Programme (Germany)	✓	✓		✓	✓	**Included**
Buildings regulation part L amendment (Ireland)		x		x		Excluded because the regulation is too new and not enough data was available
GreenerHome (Ireland)			x	x		Excluded because the MAP programme was similar and more information was available
Tuscany Biomass Regional Subsidy Programme (Italy)				x		Excluded because of limited data availability
Italian Tax Credit and VAT reduction		✓	x			Excluded because the included French tax credit is similar but has had a higher total volume.
New Energy Law (Japan)	x				✓	Excluded because policy has not met expectations
Energy Performance Requirements For New Construction (Netherlands)				x		Excluded because of limited data availability
Solar Domestic Hot Water System Agreement (Netherlands)	x		x			Excluded because of limited uptake and data availability
Household Subsidy Programme (Norway)	✓		✓	✓	✓	**Included**
Barcelona Solar Thermal Ordinance (Spain)	✓	✓	✓	✓	✓	**Included**
Technical Building code – for solar thermal (Spain)				x		Excluded because of limited data availability

Candidates	Uptake	Continuity	Uniqueness	Data Availability	Scale	Comments
Microgeneration Certification Scheme (UK)				x		Excluded because of limited data availability
Solar for London (UK)	x		x			Excluded because the programme was neither particularly unique nor did it have a particularly high uptake
Clear Skies Programme (UK)	x		x			Excluded because the project was not particularly unique, and did not have a higher uptake than the similar MAP programme
California Global Warming Solutions Act of 2006 Policy (US)		x		x		Excluded because the policy is fairly recent and therefore limited data is available
Building Technologies Programmes – tax credits and standards (US)				x		Excluded because the programme is similar in scale to MAP but less evaluation information was available
California Solar Contractor Licensing (US)				x		Excluded because of limited data availability
EPA Certification and Guidelines for Heat Pumps and Biomass Stoves (US)			x	x		Excluded because of limited data availability and because the included Solar Keymark programme is similar
Geothermal Utility Rebate Programme (Indiana, US)	✓	✓	x	✓	✓	Excluded – on original list to include, however programme administrators did not provide information. This programme is also very similar to the included Enova household subsidy programme
California Solar Rights Act (US)		✓	✓		x	Excluded because programme only indirectly supports solar installations
Hawaii Solar Water Heating for New Construction (US)			✓	x	✓	Excluded because programme has not been in existence long enough for evaluation

Candidates	Uptake	Continuity	Uniqueness	Data Availability	Scale	Comments
South Carolina Solar Heating & Cooling Personal and Tax Credit (US)			x		x	Excluded because the French tax credit programme is more significant
CombiSol Programme (European)	x	x	x	x		Excluded because the Solar Keymark (similar scope) is more significant
Climate Alliance of European Cities	✓		✓	✓	✓	**Included**
EU-CERT.HP (Europe)		✓	✓	x	✓	Originally on list to be included, however no suitable information was available
Solar Keymark (Europe)	✓	✓	✓	✓	✓	**Included**

Exhibit 15 Evaluation of candidate programmes

Selected Programmes	Policy	Sticks	Carrots	Guidance
1. Energy Action Plan of Upper Austria	✓	✓	✓	✓
2. Climate Alliance of European Cities	✓			✓
3. Market Incentive Programme (Germany)			✓	
4. Regional subsidy programmes in Austria (three in total)		✓	✓	
5. Household Subsidy Programme (Norway)			✓	✓
6. Direct tax credit (France)			✓	
7. Barcelona Solar Thermal Ordinance (Spain)		✓		✓
8. Regional promotion programmes in Austria (two in total)				✓
9. Solar Keymark (Europe)				✓

Exhibit 16 Type of instrument

Selected Programmes	Heating				Cooling	
	Solar	Biomass	Ground Source Heat Pumps	Air Source Heat Pumps	Solar	Geothermal
1. Energy Action Plan of Upper Austria	✓	✓	✓	✓		✓
2. Climate Alliance of European Cities	✓	✓	✓	✓	✓	✓
3. Market Incentive Programme (Germany)	✓	✓	✓	✓	✓[1]	✓
4. Regional subsidy programmes in Austria (three in total)	✓	✓	✓	✓		✓
5. Household Subsidy Programme (Norway)				✓		
6. Tax credit (France)	✓	✓	✓			
7. Barcelona Solar Thermal Ordinance (Spain)	✓[2]					
8. Regional promotion programmes in Austria (two in total)	✓	✓	✓		✓	✓
9. Solar Keymark (Europe)	✓				✓	

Exhibit 17 Renewable energy technologies

7.2 Programme case studies

Spar mit Solar (Austrian regional promotion programme)

Information about the *Spar mit Solar* programme was collected through a telephone interview with the programme manager, Ewald Selvicka.

Programme description

The *Spar mit Solar* programme operated within the framework of the Austrian national promotion campaign *klima:aktiv*. The *klima:aktiv* programme started in 2004 with an annual budget of €8 million and aims at supporting (among others) renewable energy technologies in Austria by promoting information, standards and quality management. *Klima:aktiv* also supported and cooperated with the states in implementing regional sub-programmes. The private company AEE INTEC, in close cooperation with the energy department of the regional government in Steiermark, designed and proposed the *Spar mit Solar* programme. It was financed by the regional government in Steiermark (40–60 per cent) and partners in the solar thermal industry (40–60 per cent), who were mainly large solar thermal technology distributors.

Because the market for solar thermal installations had stagnated in 2003/2004 (at around 30,000m^2 installed), this programme was designed to stimulate further market growth by overcoming the existing lack of general information on both the technology and the existing economic incentives provided by the state of Steiermark.

As an information campaign targeted at end-consumers, this programme clearly falls into the 'guidance' category. However, it must be seen in the context of other programmes existing or being implemented during the time the programme took place. These are, specifically, a carrot programme (financial aids provided by the state of Steiermark) and a stick programme (a solar ordinance implemented in 2008 in Steiermark.) The solar ordinance was a requirement of the national housing funding programme that provides long-term and low-interest loans to around 50 per cenr of all house builders in Steiermark. Beginning in 2008, builders of new houses are obliged to install at least 5m^2 of solar thermal collectors in order to receive these loans.

At the beginning of the programme, the campaign targeted builders and owners of detached houses because they represented the largest share of the residential market. After the solar ordinance was enacted and enforced, and solar thermal installations in new homes were provided with loans, the campaign shifted focus to the retrofit sector. The informational content of events was adjusted and, more importantly, the promotion strategy, including the slogan, was adjusted to target the retrofit sector. One campaign event specifically targeted the multi-unit sector by contacting stakeholders in the housing development sector.

This programme ran from the end of 2005 to the end of 2008. Around €100,000 was allocated annually to the programme, resulting in a total budget of around €300,000.

Context

Prior to the start of the programme, the solar thermal market in Steiermark was stagnating. One important reason for this was that the direct subsidies for this technology were relatively low in comparison to those offered in other Austrian states. The public was also not sufficiently aware of other subsidy instruments (such as the housing programme that provides loans to house builders). A lack of information on the technology in general was also thought to be hindering market growth. The government of Steiermark, therefore, decided to double the direct subsidies offered. It also decided that this action had to be accompanied by an information campaign.

The maximum subsidy (for example, for a 15m² collector providing hot water and space heating) was increased to €2000. When both incentives and tax advantages were considered, the installation of an example collector cost around €7600. Given that this collector produces 4500kWh annually, the resulting savings (in fuel) would be between €300 (wood pellets saved) and €560 (heating oil). The amortization after around 14 years is low compared to the average 20-year lifetime of the technology.[3]

The importance of this programme in the area of new buildings was decreased when the new solar ordinance was issued during the last phase of the campaign. However, the impact of the campaign in the area of retrofits remained high.

Results

This programme's primary criterion for success was the initiation of additional uptake of solar thermal technologies measured in square metres. The number of visitors to the information events would also be an indicator. The objective set was to end the stagnation on the solar thermal market and to eventually increase the installed collector area by 10 per cent per year, meaning a total impact of 30 per cent market growth during the project duration. This target was more than achieved: from 2005 on, the market grew by 30 per cent each year, meaning that the solar thermal market in Steiermark almost doubled during the programme duration. Around 1600 interested end-consumers visited the events per year. This clearly shows the success of the campaign. In addition, the Austrian solar thermal industry, represented by the national industry association (Austria Solar), described the programme as 'very successful' in official statements and the state of Steiermark expressed its interest to continue the campaign.

The exhibit below clearly shows the stagnation of the solar thermal market between 2000 and 2004. During the project duration, market growth rates increased significantly.

The influence of the programme on market growth must be rated as very high, especially during the first two years. It is believed that the programme provided the primary factor (information) to end-consumers before secondary factors (such as high gas prices and financial incentives) could become effective.

Design Process

The framework for the programme was provided by the national *klima:aktiv* programme supporting regional solar thermal campaigns. The project idea was developed by AEE INTEC as a private company with significant experience in designing similar projects and well-established connections within the renewable energy and solar thermal industry sector. The programme was designed in close cooperation with the relevant (semi-)public bodies such as the energy department of the regional government and the regional energy agency.

The main barrier to the enhanced deployment of solar thermal technologies in Steiermark was believed to be the lack of information available to the end consumer. Both general information on solar thermal applications and specific information on financial aid opportunities had to be provided to the broad public and confidence in the technology had to be created.

The information campaign was combined with the doubling of the incentive provided by the government. The alternative to this combination would have been an even larger increase of subsidies without accompanying promotion. The chosen combination achieved the targets more economically.

Given the identified information barrier, a number of information events were planned and the cooperation with regional media had to be implemented.

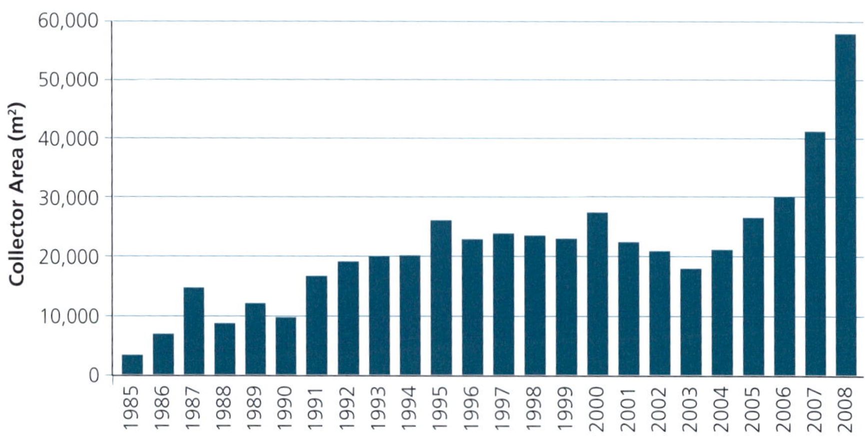

Exhibit 18 Annual new installation of solar thermal collectors (m²) in Steiermark[4]

The information was given to the target groups by independent experts in order to gain the confidence of end-consumers.

Implementation process

AEE INTEC was responsible for the implementation of the promotion campaign. A professional media design partner provided the promotion materials. Cooperation with a regional print medium was established. This newspaper group is represented within the region of Steiermark by a number of subregional affiliated media. The advertising for and coverage of the events were done by these subregional newspapers.

Eight to ten information events were organized per year, approximately one per subregion, thereby covering the whole of Steiermark. The events took place in subregional central towns, attracting visitors from the surrounding rural areas. During the information events, only independent experts were in contact with visitors while the industrial partners stayed in the background. After the event, the industrial sponsors could come in contact with potential customers by presenting their companies in an attached exhibition.

In addition, the campaign was advertised with information stands at several exhibition events related to the building sector.

Evaluation process

Monitoring of programme success in the short and medium-term was done by counting the number of visitors to the events. Long-term monitoring was done by determining the increase of installed collector area. A general (formal) oversight was provided by the relevant public authorities and, more informally, by the co-financing industry partners. Progress was reported in meetings with the financing parties as well as in the interim and a final report.

The government of Steiermark evaluated the project both at an interim point and at the end of the project.

Challenges and barriers

One problem encountered was that the interest of the media in general was initially not sufficient. This was overcome by establishing continuous contacts with the media and repeatedly providing information to newspapers and TV. The media interest was increased significantly during the project duration.

Programme conclusions

The programme contact recommended that this information campaign be replicated in regions where the lack of information and confidence are the main barriers to increased uptake of solar thermal technologies in the residential sector. The basic design of the campaign could also be applied to other technologies. Suitable partners (project management, design partners, media partners) would have to be identified and funding would need to be provided. This could be done by public bodies, however, the cooperation with industry partners is an advantage because the contact between customers and businesses results in actual market growth. Relevant businesses should be active in the target region or companies from outside the region need to be convinced to get involved.

The programme contact also stated that the concept of a roadshow with several information campaigns covering a whole region appears to be a very useful method of reaching the broad public. It is especially well-suited for promoting a new or altered subsidy scheme that needs to be communicated to interested consumers. This way the impact of subsidy schemes can be increased significantly with relatively low costs.

Analysis

This appears to have been a very successful programme. The pairing of an information campaign with an increased subsidy seems to have created significant market growth. The existence of the national *klima:aktiv* programme may have been another factor contributing to the success of this regional programme. The development of partnerships with media and industry also appear to be factors in the programme's success. Availability of suitable systems and an adequate installation and maintenance infrastructure do not appear to have presented barriers, presumably because the market was adequately mature when the programme began.

Key lessons learned

The *Spar mit Solar* programme provides the following lessons:

- Programme design can be effectively done by private companies with specific expertise.
- Close cooperation with relevant public bodies and the industry needs to be ensured.
- Campaigns with long duration need to be designed in a way that provides the flexibility to react to changes in the market/policy context.
- Media coverage is crucial to the success of information campaigns.
- Detailed planning of media cooperation must be part of the design process.
- Promotion materials must be of good quality (for example, involving professional design partners).
- Information campaigns must be focused on a limited set of technologies.

- Business interests should be kept in the background. However, the contact between end-consumers and the businesses involved has to be separately ensured.
- In order to cover the whole region, a roadshow should include several events in alternating locations. The events should be located in central towns, not in remote rural regions.

Ja zu Solar (Austrian regional promotion programme)

Information about the *Ja zu Solar* programme was collected through a telephone interview with Mr Oberhuber, the director of the regional energy agency in Tirol, who was responsible for programme implementation.

Programme description

The local regional energy agency in Tirol (*Energie Tirol*) started *Ja zu Solar* independently in 2004, coinciding with the start of the national Austrian *klima:aktiv* campaign. Because *klima:aktiv* was mandated to cooperate with the Austrian states, *klima:aktiv* provided some of the initial funding for *Ja zu Solar*. Additional funding was provided by Tirol industrial partners (collector manufacturers). In the first year, 2005, the budget of €80,000 was needed for initiating the campaign, for example, for contracting a professional communication agency that designed promotion materials and the slogan. *Klima:aktiv* also contributed, the first year, by providing information material and speakers for regional information events. In subsequent years, an annual budget of €30,000–40,000 has been sufficient.

After the programme was designed and funding was ensured, *Energie Tirol* proposed the campaign to the responsible governmental bodies and received approval for three years (2005–2007). A programme target of increasing the installed collector area from 160,000m² in 2005 to 300,000m² in 2010 was set. Since a lack of information on the consumer side was identified as one of the main barriers to increased uptake of solar thermal technologies, the implementation of a broad information campaign was seen as an appropriate tool to reach this goal.

The information campaign was launched together with a special direct subsidy for solar thermal installations and the programme also coincided with a rise of prices for conventional fuels for household heating. These three factors were responsible for a boom of solar thermal technologies during the following years.

Two target groups were addressed specifically by this programme:

1 Owners and builders of detached houses (both new and existing buildings) were addressed by a number of information events (around 15 per year) and

by the distribution of materials such as information folders. Around 150,000 households (around 60 per cent) were reached through these means.
2 The tourism sector (primarily hotel owners) was also targeted by individually contacting thousands of tourism businesses and organizing special events.

Context

Around 16,000m^2 of solar thermal collectors were installed with the help of subsidies in 2004. In order to increase market growth, a special direct subsidy and an information campaign were implemented. These actions, together with a sharp increase in prices for conventional fuels, led to a boom on the solar thermal market during the following years: In 2005, 30,000m^2 were subsidized and in each of 2006 and 2007, 90,000m^2 were subsidized.

In spite of high oil and gas prices, calculations indicate that the installation of solar thermal collectors amortizes only after more than ten years, even when including subsidies. Solar thermal technologies are therefore not attractive from a financial point of view. In many cases, people visiting the programme events were interested in the technology from the beginning. However, these people often were unsure about the viability of the technology and a general confusion concerning the technology had to be addressed.

Results

This programme's primary criterion for success was the initiation of additional uptake of solar thermal technologies measured in square metres installed. The number of visitors to the information events would be an indicator as well. For the financing industrial partners, the number of additional business opportunities resulting from their presence at events was another way to measure success.

The target set was to increase the installed collector area from 160,000m^2 (2005) to 300,000m^2 within five years. In 2008, a total of more than 400,000m^2 had been installed in Tirol. The boom of the market in 2006 and 2007 was unexpected so that the target set seems too low today.

With almost 10,000 visitors to information campaigns, 150,000 households reached with information materials and broad coverage of the topic within regional media, the impact of this campaign must be rated as very high. However, its impact on the market growth cannot be directly measured.

Design process

The design process was initiated by the regional energy agency *Energie Tirol*. With the help of the national *klima:aktiv* programme, the campaign, including the funding scheme, was set up and proposed to the responsible governmental bodies. Technology manufacturers financing the project were also included in the design process.

Implementation process

Around 15 information events with around 250 visitors each were organized annually. The financing industrial partners were allowed to present their companies after the events in order to reach potential customers. They were kept in the background during the information event and their presentation was strictly limited to flyers and a poster. It was seen as very important to organize information events, not sales events.

Promotion materials were provided by a professional communication company and *Energie Tirol* cooperated closely with the municipalities where the events were planned. The municipalities were provided with promotional materials that included the emblem of their municipality. Having the municipalities distributing material such as the invitations to the events, and thereby winning the consumers' confidence, contributed significantly to the success of the programme.

After the first year, additional features were implemented. Local installers were included in the events as sponsors and as presenters of their companies. In addition, energy advisors were sent to the municipalities two or three weeks after the events in order to give advice to the inhabitants individually. This was done by going from door to door. In this manner, a large number of households were reached directly. This action also allowed more specific topics, such as quality issues or the architectural integration of solar thermal installations, to be addressed. In addition, standardized bidding documents were distributed and potential customers were advised on how to find the right installer.

The campaign was also very successful in addressing the tourism sector. The strategy was to convince potential users of the positive image for the tourism sector that might be created by the application of sustainable energy technologies. Information on financial incentives for this sector was provided. The possible accumulation of incentives in this sector made the installation of solar thermal technologies very feasible.

Evaluation process

This programme has not been formally evaluated.

Challenges and barriers

No specific barriers or challenges were identified.

Programme conclusions

The basic design of this campaign is very well-suited for the distribution of general information on renewable heating technologies. A large number of households were provided with advice concerning solar thermal technologies.

Consumer uncertainties were addressed and knowledge about financial assistance and on quality and efficiency issues was spread.

The support from the municipalities was crucial to the success of this campaign. Cooperative and interested municipalities would be required in order to implement this campaign in other regions.

Analysis

This programme appears to have been very successful, although it could have been improved by adding a formal evaluation component. Cooperation with a national programme, industry partners and municipalities seem to have been factors in the programme's success. Another factor appears to be the pairing an information campaign with a subsidy programme.

Key lessons learned

This programme provides the following lessons:

- Regional energy agencies are well-suited for designing and implementing similar campaigns.
- Financial capacities of regional/local industrial partners should be used for co-financing.
- Close cooperation with relevant public bodies and industry needs to be ensured.
- The support from municipalities significantly increases the impact of such campaigns.
- Advertising for events is not necessary: coverage through regular newspaper reporting is sufficient.
- Promotion materials must be of good quality (for example, involving professional design partners).
- Business interests should be kept in the background. However, contact between end-consumers and the businesses involved has to be (separately) ensured.
- Installers should be addressed by the events. Information on technologies (quality and efficiency) is very useful to them.
- Installers can also be included as sponsoring partners. Direct contact between consumers and installers is a good way to promote market development.
- Quality, efficiency and architectural issues should be addressed. Installations with poor performance or appearance can be avoided this way.

Salzburg (Austrian regional subsidy programme)

Information about the regional subsidy scheme in Salzburg was collected through a telephone interview with Rudolf Krugluger, the responsible head of unit in the regional administration.

Programme description

The state of Salzburg provides low-interest and long-term loans to homeowners for new houses and for renovation. These loans are only granted under the following conditions: the house owner's income is below a certain threshold and either the house owner provides a bank guarantee or the state of Salzburg is included in the register of real estate for the respective building. When homeowners install REHC technologies, the loan amount is increased. In many cases, the homeowner's income is too high or they are reluctant to provide bank guarantees or to agree with the change in the register of real estate. Therefore, only about 50 per cent of the new buildings are built with the help of state loans.

During the 1980s, the regional government of Salzburg was already aware of the fact that, in addition to the loan programme described above, a direct subsidy was necessary to promote solar thermal technologies on a large scale. Although the subsidy programme began with solar thermal technologies, biomass heating technologies were included in this scheme in the 1990s, and heat pumps were added in 2009. The connection to existing district heating grids (providing heat from renewable sources) is subsidized as well.

The subsidy scheme is designed and implemented by the regional government administration. The regional government finances the subsidies and decides annually on the budget for the next year.

In 2009, the total amount of the subsidy is limited to a maximum of 30 per cent of the total investment. For the installation of a pellet boiler, for example, €1000 can be paid as a base sum. Additional subsidies are paid if:

- High efficiency requirements are met (€400).
- Buffer storage tanks are installed (€500).
- Hydraulic adjustment is performed (€200).
- High efficiency pumps are installed (€50 per pump).
- Particle removal is provided (€500).
- An energy audit is performed according to the guidelines, which is required for biomass heating (€200).
- The heating system inspection is performed according to the guidelines (€100).
- The building is well insulated (€100–1000).
- A heat recovery system is installed (€300–500).
- The boiler is combined with solar thermal (€500).
- The heating system is changed from fossil to biomass fuels (€500).

These very detailed requirements make the investment in additional efficiency measures more attractive to the end-consumers. Since the multi-unit dwelling sector is better served with the state loan programme, this scheme mainly targets the detached house sector (new buildings and renovations).

The installed per capita capacity of renewable heating technologies is rather small in Salzburg when compared to other Austrian states. The initial reason for implementing this subsidy scheme was to promote solar thermal market growth. This has not worked, mainly because the amount of subsidies was and still is small compared to other similar schemes in other Austrian states. The number of applications per capita for subsidies for solar thermal installations for example is one of the lowest among the Austrian regions.[5] Today, the promotion of market growth is not the primary target of this scheme. Due to limited budgets, this programme mainly aims at promoting efficiency and quality of the technologies and the installation services.

Context

This subsidy scheme is not specifically designed to promote significant market growth. The installed per capita capacity of renewable heating technologies is small in Salzburg when compared to other Austrian states, which may be due to the comparably low subsidy amounts. These lower subsidy amounts mean this scheme is unlikely to convince otherwise uninterested homeowners to invest in renewable energy technologies. Instead, the scheme targets people who are thinking about installing renewable heating anyway. The subsidy then ensures that these people have a slight financial advantage compared to conventional heating technologies and that the installations are of high efficiency and quality.

Results

Today, the programme's success is not measured in terms of market growth achieved. The aim is to promote the installation of highly efficient technologies. A unique feature of this programme is a detailed catalogue of technical specifications that have to be met in order to receive the subsidy. It can be assumed that the quality of installations in Salzburg is very high compared to other regions, however, no data on this is available.

Another goal of the programme is to approve as many applications as possible with the given budget and with low personnel costs. In 2008, more than 1000 applications were approved by three administrators.

Design and implementation process

Because the programme began in the 1980s, little is known about the original design and implementation process. Now, the government annually determines the budget and the head of the programme drafts the directive including the amount of subsidy provided to each of the technologies. These draft directives then have to be approved by the government.

The continuous redesigning of the programme is part of the implementation process. The aim of ensuring high quality and efficiency of the subsidized installations requires continuous adaptation of the technical requirements and application procedures necessary. In general, the technical specifications have to keep up with the state-of-the-art of the respective technologies. The application process has also evolved: originally, the applicant simply turned in the invoice for the installation and received the subsidy. When examining the installations afterwards, the administration often found the subsidized installations had significant flaws. Now, the installer has to turn in a plan for the installation that allows major quality issues to be checked before the installation is approved.

The second major aim of the design of the scheme was the cost-efficient handling of applications. The processing of applications had to be automated as much as possible. Today, there is a dedicated webpage presenting information on the subsidy scheme and allowing for online application.[6] All steps during the application process are automated and can be carried out online. As a result, three administrators were able to handle more than 1000 applications in 2008. The software for application handling is updated and adapted monthly in order to get rid of flaws and to ensure maximum user-friendliness.

The programme was designed to target builders and owners of detached houses. Renovations and new buildings are treated equally. The amount of subsidy available per square metre decreases with the increasing amount of area installed per building, so owners of multi-unit dwellings are far better served with the state loan programme.

Evaluation process

General oversight for the scheme is provided from within the hierarchy of the government administration. No regular evaluation is done. The last evaluation of the programme was done in 2004 by an external evaluator, the regional energy agency.

Challenges and barriers

The challenge of improving the data handling system has to be met continuously. Applicants and installers dealing with the system provide feedback on bugs, which are then eliminated during monthly software revisions. Another aim of the software revisions is the simplification of the application process in order to optimize user-friendliness.

The programme also had to deal with a lack of information on the end-consumer side. People are often not aware of the opportunities provided by this scheme, they also often distrust the intention of the administration. For example, an energy certificate, based on an energy audit, is necessary for receiving a subsidy. People are reluctant to get the energy certificate because they fear that a low mark might cause higher taxes or other penalties. Programme staff try to inform potential applicants by presenting the scheme at events such as building-related fairs, but the lack of an accompanying information campaign might be seen as a weakness of this programme.

Programme conclusions

The very basic design of the scheme has already been applied in a number of regions and countries and could be replicated elsewhere.

Specific features that might be considered as useful to other regions/countries are:

- The highly sophisticated and well-developed online application tool. The experience gained during developing this tool might be useful for other programmes interested in improving the cost-effectiveness of application handling.
- The detailed specification of technologies and installations that qualify for subsidies. This might be criticized as being too complicated, but a detailed catalogue of technology specifications ensures high quality and efficiency.
- The requirement that a plan of the proposed installation be provided by installer. This avoids basic mistakes.
- The requirement for an energy certificate (based on an energy audit) as a prerequisite for approval. As a result, the concept of energy certificates is very well-known and frequently used in Salzburg.
- The establishment of a 'high efficiency package'. For example, when a solar thermal installation meets a number of additional requirements (such as minimum storage capacity, certain technological threshold values, optimum collector orientation, insulation of tubes) the applicant can receive an additional €600. Most applicants ask their installers to plan the installation in a way that meets these efficiency requirements. As a result, most of the installations subsidized are state-of-the-art.

Analysis

This programme is interesting in that it measures success in terms of high quality installations and efficient application processing. Increasing the number of installations does not appear to be a goal.

Given the programme goals, the long programme history and the continuous tweaking of the system requirements and the application process appear to be the key factors in the programme's success. The establishment of bonuses for high efficiency installations appears to provide a carrot for encouraging households to install better systems than they may have originally planned.

It appears that an accompanying information campaign and partnership with industry would increase the impact of this programme.

Key lessons learned

The following lessons can be learned from this programme:

- Additional programmes are required to reach households not eligible for primary programmes. In this case, the direct subsidies are targeting households not eligible for the state loan programme.
- When subsidies are comparably low, applicants will primarily be people who are already interested in the technology.
- Given the proper tools and programme set-up, a large number of applications can be handled with low personnel costs and without administrative barriers.
- In order to ensure a high quality and efficiency of the installations, continuous adjustment of technical requirements is necessary as the state-of-the-art of the market develops.
- Very detailed technical requirements initially seem to be too complicated, but because only the installers have to deal with these specifications, they pose no barrier to the applicant.
- Demanding a technical drawing from the installer before approving the application is a useful tool for preventing basic mistakes during installation.
- Demanding an energy certificate from the applicant was shown to be a feasible way to increase the use of these certificates.
- The process of application handling can be automated to a large extent. However, the system has to be improved continuously using feedback from users.
- An accompanying information campaign might increase the impact of this programme.
- Closer cooperation with installers might increase the impact and to prevent mistakes during planning and the application process.

Vorarlberg (Austrian regional subsidy programme)

Programme information about the regional subsidy scheme in Vorarlberg was collected through a telephone interview with Wilhelm Schlader from the *Energieinstitut Vorarlberg* and Christian Vögel from the Vorarlberg state administration.

Programme description

The energy strategy for the residential sector includes two types of financial incentives provided to households for the installation of REHC technologies: the housing programme and a direct subsidy programme.

The housing programme was most recently amended by the Vorarlberg government in 2009 for the period 2009/2010. The programme provides low-interest loans for building or renovating houses. In case of minor renovations (such as a change of heating system), a direct subsidy can be granted. As in other Austrian states, the programme sets eligibility limits regarding the applicants' incomes and the total costs of the building projects. Furthermore, the houses have to meet minimum requirements on insulation (low heating demand) and innovative heating systems have to be installed. These systems can be district heating (biomass derived) or highly efficient gas, biomass or heat pump technologies. If the building is not connected to a district heating system, the main heating system has to be combined with a solar thermal energy system.

Despite these strict requirements, an unusually high share (more than 70 per cent) of the new houses in Vorarlberg are constructed with the help of this loan programme (for example, 466 detached houses in 2008). A similar uptake has been achieved with renovation projects. The main reasons for this high acceptance rate are the very attractive loan conditions and the recently implemented easing of income thresholds. The amount of loan granted can be increased by collecting 'eco-points' based on the energy efficiency measures installed. The loan can amount to €1150 per square metre of floor space. Projects with very high quality (many eco-points) systems are exempt from the income limits, providing access to the programme for high-income households.

Although solar thermal technologies and innovative heating are already required in the housing programme, the state of Vorarlberg provides additional direct subsidies for heat pumps, biomass heating and solar thermal installations. These subsidies do not depend on income but rather require high efficiency and high quality of installations. This means that REHC installations in new houses are funded twice (loans and subsidies) in many cases. For renovation projects, on the other hand, a biomass boiler would bring additional eco-points, but the investment is not added to the total project cost which therefore limits the loan and subsidy amounts.

Both the loan and the direct subsidy programmes are designed and financed by the state. The 2009 budget for the housing programme is €92 million for loans for new buildings and €33 million for renovations. An additional €3 million is dedicated to subsidies for the installation of solar technologies (around 1000 cases), around €1 million for biomass (500–600 cases) and €700,000 for heat pumps (around 400 cases per year).

Budget decisions are based upon estimates of the expected number of applications. If the budget is exhausted during the year, applications are handled in next year's budget or the budget is extended. This provides security to homeowners and gives the programme continuity.

The administration is responsible for the implementation of both programmes including the processing and approval of applications. However, the annual adaptation of programme designs (subsidy amounts, technical requirements, quality control) is done in close cooperation with regional organizations such as *Energieinstitut Vorarlberg*, who provide expertise on technologies and markets.

Solar thermal applications have been promoted with subsidies since 1991. The initial focus of the programme was to enhance market growth. Today, the market penetration of solar thermal technologies and of heat pumps is very high in Vorarlberg. Solar thermal collectors are installed on the largest part of new residential buildings and heat pumps are the main heating system in around 63 per cent of the new buildings. Although the technologies are installed because of the requirements of the housing programme, the subsidy programme ensures high quality and efficient installations because end-consumers try to get additional funds by demanding installers meet the strict technical requirements of the subsidy programme. While the technical requirements have been periodically raised, the amount of subsidies has remained rather stable over the past years, resulting in comparatively low amounts that still are sufficient to promote highly efficient installations. Examples of subsidy amounts for detached houses are:

- €1600 for ground sourced heat pumps (vertical collector).
- €1200 for other ground or water-sourced heat pumps.
- A maximum of €1900 for solar thermal (water heating only).
- A maximum of €3700 for solar thermal (water and space heating).

For multi-unit dwelling, the total subsidy amounts can be very high and REHC technologies become increasingly attractive in this sector.

Although the technical requirements are strict and detailed, they do not hinder the uptake of this programme. Before the application, the applicant only has to provide an energy certificate. The conformity of the installation with the technical requirements is inspected randomly after the installation.

The programmes equally target all relevant sectors: new buildings and renovations, detached houses and multi-unit dwellings.

Context

The uptake of REHC technologies in the residential sector is very well-developed in Vorarlberg. More than 200 applications for subsidies for solar thermal installations were registered in the first quarter of 2009, which is, per capita, among the highest numbers among Austrian states. In 2007, 13,000 solar thermal installations were counted in Vorarlberg and the collector area per capita was $0.44m^2$. With more than 60 per cent of the new houses equipped with heat pumps and an even larger share equipped with solar collectors, the uptake of REHC in this region is remarkably high.

The programmes target renovations as well and for the multi-unit dwelling sector, the direct subsidies can be very attractive: For automated pellet boilers, for example, there is a basic €1700 subsidy per building plus €700 per flat. Given a group of multi-unit buildings, the installation of micro-grid with a central heating unit can become very feasible even with the limitation of the total subsidy amount to 30 per cent of the investment costs.

Results

The results of the programmes can be measured in terms of collector area installed or as the share of houses equipped with REHC technologies. As a result of the programmes, the market penetration of REHC is very high in Vorarlberg. The success of the programme is also expressed by the high uptake of the programmes, specifically in the number of applications and total subsidies paid, as seen in Exhibit 19 and Exhibit 20.

One of the major aims of the direct subsidy programme is to ensure the quality of installed REHC technologies. The success concerning this aim is not easy to measure. On-site inspections of installations show that most of them meet the programme requirements. Given the fact that these requirements are comparatively high, it is most probable that the quality of installations in Vorarlberg is relatively high.

Design and implementation process

The continuous redesign of the programme is part of the implementation process. Both the housing programme and the subsidy programme are adapted continuously by the programme administration in close cooperation with independent organizations.

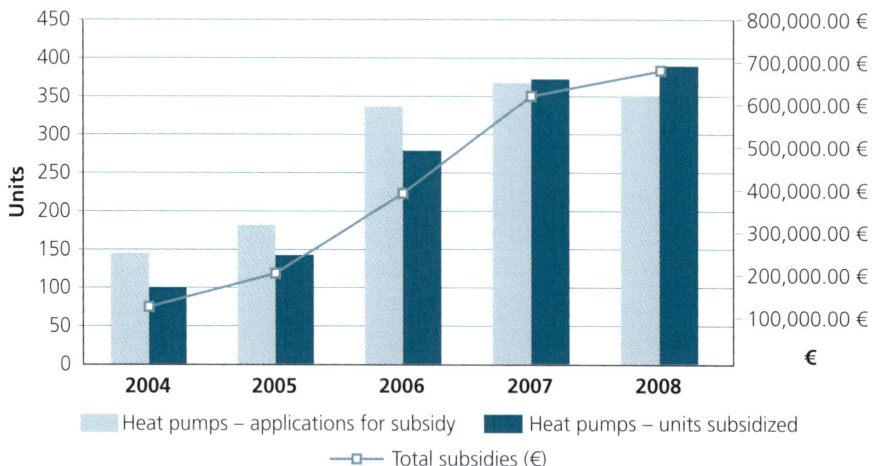

Exhibit 19 Heat pump applications made, units installed, subsidies in Vorarlberg

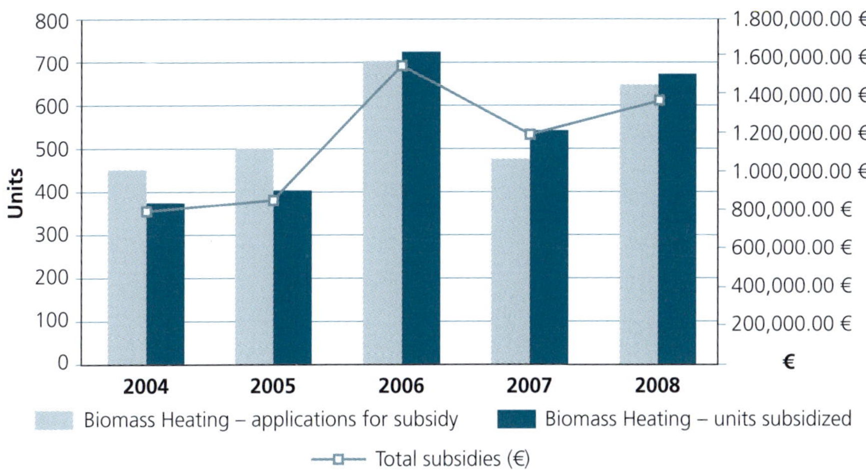

Exhibit 20 Biomass applications made, units installed, subsidies in Vorarlberg

The strategy behind the adaptation process is to provide continuity. The lifetime of the programmes is not limited and the budgets are high enough to guarantee funding throughout the year. Budget shortages at the end of the year are dealt by handling surplus applications under the next year's budget or by extending the budget. The directives are amended in small steps with an ever-increasing demand on quality and efficiency while the subsidy amount has been kept stable during the last years.

The application process is kept deliberately simple. The applicant provides an energy certificate for the building, has the installation done and then submits the application together with relevant invoices and the installer's affirmation of appropriate installation. Depending on the technology, additional documents concerning technical features can be required. The approved installations are inspected randomly after installation.

Evaluation process

General oversight of the programme is provided within the administration. Basic programme performance indicators are the collector area installed or the number of installations compared to the resulting programme costs.

Energieinstitut Vorarlberg provides in-depth analyses of programme statistics. Furthermore, measurement programmes (such as measuring the efficiency of heat pumps) and on-site controls of approved installations are performed. This data shows for which sector or technology better support is needed and allows for the identification of quality problems. The results can then be used for the regular amendments of the programme directives.

Challenges and barriers

No specific barriers or problems were identified.

Programme conclusions

The direct subsidy programme is very suitable for regions or countries where:

- The uptake of REHC is already high.
- The use of REHC is required by an accompanying loan programme and where the uptake of the loan programme is very high.
- The use of REHC is required by building codes or other ordinances.

In these cases, the use of REHC in new houses or during renovation is already high. An additional subsidy programme can then contribute to quality improvement with relatively low costs.

Analysis

This programme is well-suited to the unique situation in Vorarlberg where the installation of REHC systems is often required and the market is well-developed. By providing additional subsidies for high-quality installations, the overall quality of the installations is raised.

Close cooperation with other organizations, stable funding, simple applications and continuous redesigning all appear to be important factors in the programme's success.

Key lessons learned

This programme provides the following lessons:

- A housing loan programme with high uptake can have a similar effect as the required use (through building codes, ordinances) of REHC.
- Factors increasing the uptake of housing programmes can be:
  - attractive loan conditions;
  - income thresholds not set too low;
  - a direct subsidy available instead of a loan, especially for smaller renovation projects (such as the replacement of a heating system).
- High-income households can be addressed by increasing the income threshold for high-quality projects.
- An additional subsidy programme can be used to ensure quality and efficiency of REHC installations. A large impact on quality can then be achieved with comparably low budgets.

- In order to ensure high quality efficient installations, continuous adjustment of technical requirements is necessary as the state-of-the-art of the market develops.
- The application and approval process can be kept very simple without the risk of decreasing quality of installations.
- Demanding an energy certificate from the applicant was shown to be a feasible way to increase the use of these certificates.
- Decisions on programme design should not exclusively be done by the administration. The technical and market expertise of other organizations should be used.

Upper Austria's Energy Action Plans

Information for the Upper Austria solar thermal and small-scale biomass policy was gathered from the following sources:

- *Umsetzung des O. Ö. Energiekonzeptes* (implementation reports of the Upper Austrian Energy Action Plan, published annually for the previous year).
- *Energieförderung in Oberösterreich* (Stand 3/2009).
- Christiane Egger, *Solar Upper Austria – A Shining Example for Europe*, Linz, 2004.
- Christiane Egger, *Systemic Market Promotion in One of Europe's Leading Solar Thermal Regions*, Solar Thermal Upper Austria. Presentation held on 30 May 2009 at the ProSTO expert workshop in Brussels.

Programme description

Upper Austria is one of nine federal states of Austria and is the clear leader when it comes to the use of renewable energies. The state has a long tradition of supporting small-scale renewable energy installations (solar thermal and biomass) as well as the efficient use of energy, especially in buildings.

The key to Upper Austria's success in renewable energies lies in the integrated Energy Action Plans, which have guided its renewable energy policy over the past 16 years. The first one was approved in 1993, setting targets for the year 2000. In 2000, the second edition, dubbed 'Energy 21' was launched, setting targets for 2010. In October 2007, the government of Upper Austria approved the latest programme, 'Energy Future 2030', which outlines the path for the next two decades. This programme foresees a 39 per cent reduction of the heat demand by 2030, which, together with the increased use of renewable energies, will lead to 100 per cent of the heating demand being covered by renewable energy sources.

The Energy Action Plans do not focus on a single market barrier or technology, but provide a well-defined mix of support mechanisms, which have led to a strong development in the use of (small-scale) biomass and solar thermal technologies.

Each plan consists of:

- Clear, ambitious, but achievable targets.
- Measures addressing the various barriers to growth.
- Financial incentives.
- Information and public awareness-raising.
- (Independent) advice on energy issues.
- Legal/regulatory measures, including a de facto obligation for the use of renewable heating in new buildings (a requirement in the housing subsidy programme).
- Development of new business models for the sale of renewable energy technologies, especially in the form of contracting models.
- Training of professionals.
- Support of regional R&D and research institutes.

While the concrete details of the support schemes have been adapted numerous times, the overview of the current situation provides a good example of the overall level of support:

Solar thermal systems are publicly supported with a grant of €1100 per system and €100 per square metre of flat plate collector area (€140 per square metre in the case of vacuum tube collectors), up to a maximum of €3800. To discourage lower-quality products, collectors which are not Solar Keymark certified receive a lower grant. To ensure the good working of the system, the incentive programme requires the installation of a heat meter.

A financial incentive of 30 per cent of the net investment cost is offered for wood pellet and wood chip boilers (up to €2200 per installation). Additional bonus payments are granted when boilers of more than 15 years of age or heating oil or liquid gas installations are replaced. Each such installation may receive up to €3700 in financial incentives.

In 2008, a de facto solar obligation was introduced for new buildings by making solar a requirement for receiving the popular housing subsidy for new residential buildings. As of 2009, this obligation has been expanded to renewable energies in general.

Context

The different Energy Action Plans of Upper Austria have been influenced by an increasingly accepted need to decrease the use of fossil fuels to combat climate change and the increasing awareness of the potential for economic growth through renewable energies and energy efficiency. Building an industry which offers energy equipment for the future is seen as a wise strategy to help create long-term sustainable jobs in Upper Austria. To this end, a network of 150 green energy businesses ('*Ökoenergie* Cluster') has been established, which fosters cooperation between the partners with common training, information, research and export activities.

Results

The results of the very long-term energy policy of Upper Austria are remarkable. During the first Energy Action Plan, from 1994–1999, the share of renewable energy in the overall energy mix of Upper Austria was increased from 25 per cent to 30 per cent. Today, more than 40 per cent of the heat demand is met by renewable energy. Consequently, in 1999–2007, oil consumption for heating has been reduced from 36 per cent to less than 1 per cent of the heating sector load.

Biomass achievements up to this point include:

- More than 30 per cent of new one-family houses are equipped with modern biomass heating systems.
- 30 per cent of municipalities use biomass as their main heating resource.
- Biomass provides 14 per cent of total energy consumption from more than 30,000 biomass heating installations.
- Steady growth has lead to more than 1600MW capacity in 2007.

Solar thermal achievements up to this point include:

- Highest per capita solar collector area/inhabitant in Austria and Europe: $0.7m^2$ of collector area installed per inhabitant, compared with $0.05m^2$ per inhabitant for the whole of Europe.
- Around $900,000m^2$ (600MW) of installed solar thermal capacity.

Design process

The Energy Action Plans are largely developed by the *O. Ö. Energiesparverband*, the energy agency of Upper Austria, which is also responsible for most of the implementation measures. The plans are approved by the regional government of Upper Austria. By closely involving the energy agency in charge of most of the implementation, know-how from the previous plans and implementation is fed-back into the development of each new plan.

Implementation process

The implementation of the Energy Action Plans is largely coordinated and carried out by the *O. Ö. Energiesparverband*. It was set up by the regional government in 1991 to promote energy efficiency, renewable energy sources and innovative energy technologies. The main target groups are private households, public bodies (such as municipalities) and businesses. The energy agency is active on local, regional, national, EU and international levels with numerous projects and programmes.

Among other things, the *O. Ö. Energiesparverband*:

- Provides energy information and helps raise public awareness through targeted campaigns.
- Provides specific energy advice for households, public bodies and businesses.
- Develops and promotes sustainable building programmes.
- Offers training to professionals in the energy sector.
- Manages the green energy business network '*Ökoenergie* Cluster (OEC)'.
- Manages the regional energy R&D programme.
- Promotes energy performance contracting, especially for public buildings.
- Supports municipalities in developing local energy strategies

Evaluation process

The implementation of the Energy Action Plans is monitored regularly, although not by an independent organization but by the *O. Ö. Energiesparverband*, which is also responsible for most of the implementation. The government publishes an annual implementation report, which is prepared by the *O. Ö. Energiesparverband*. These reports provide detailed data on energy usage, the numbers of renewable energy installations and their energy production. They also report on the measures taken to further decrease the energy demand and to increase the use of renewable energies. However, a critical analysis of individual implementation measures, their effectiveness and possibly even efficiency is largely missing in these reports.

Challenges and barriers

The Energy Action Plans try to address the various challenges and barriers through targeted implementation measures. Upper Austria follows a step-wise approach, which aims at extending renewable heating and cooling to more applications, building types and commercial sectors. While the typical one and two-family houses are already well-covered by the implementation measures (information, advice, grants), other sectors and market segments need special attention and strategies. For example, larger solar thermal installations in residential apartment blocks need better trained planners and installers, and the buildings' owners, often large housing companies, are approaching investment decisions differently from the private homeowners. In the next phase the *O. Ö. Energiesparverband* aims at specifically targeting industrial companies when promoting solar thermal solution.

Programme conclusions

The single most important factor in the success of the Upper Austrian Energy Action Plans lies in their continuity. For solar thermal, for example, financial incentives have been offered continuously since the late 1970s. With strong political backing and the agreement of longer-term targets in the Energy Action Plans, Upper Austria has created a stable support framework that has resulted in positive investment climate. The stop and go support so often seen in other regions and countries was avoided and the renewable heating sector is able to trust that support will not suddenly end in Upper Austria.

The mix of measures used by the State of Upper Austria has proven highly successful in stimulating the demand for renewable heating and cooling solutions. In carrying out the Energy Action Plans, the regional energy agency relies on three pillars: carrots (such as direct grants for the installation of renewable energy systems), sticks (such as new renewables obligation in new buildings) and guidance (such as independent energy advice given to end consumers).

The implementation addresses the various barriers to growth and aims at a step-wise build-up of a self-intensifying market; both on the demand side and on the supply side. While the former is addressed mostly through information campaigns, independent energy advice and direct grants, the latter is being addressed through the Network of Green Energy Businesses ('Ökoenergie Cluster'), the training of professionals and the support for energy R&D.

Underlying this mix of measures is the experience that the market only grows as fast as its slowest participants (manufacturers, planners, installers, consumers). This approach therefore avoids the sometimes one-sided approaches seen in other programmes, which focus only one or two barriers to growth, but neglect other important growth factors.

Analysis

The Upper Austria renewable energy policy has been very effective. This appears to be the result of the following factors:

- A long-term, continuous commitment.
- Clear, ambitious and achievable goals.
- A full complement of measures designed to achieve the policy goals.
- Recognition that all market participants must be supported.

The Upper Austria renewable energy policy stands in contrast to the 1997 New Energy Law in Japan which was intended to reduce oil dependency by encouraging the use of alternative energy resources. This policy was also supported by a wide range of programmes but was ineffective in reviving Japan's solar market because it could not overcome the significant decrease in price of conventional heating technology.[7] This suggests that changes in fuel prices and other externalities also are critical to the success of a policy.

Key lessons learned

The Upper Austria's Energy Action Plans suggests the following key lessons learned:

- Long-term targets and strategies help avoid the stop and go support seen in many other support programmes. The continuity in Upper Austria has created a favourable climate for the development of a thriving renewable energy sector.
- By addressing the different market barriers through separate measures, Upper Austria avoids having a the slow development of one group of market participants (manufacturers, planners, installers, consumers) become a bottleneck for the overall development of the renewable heating market.
- Given the continued strong backing of the government, a regional energy agency can become a key driver for the development of renewable energies.

Umweltlandesfonds Steiermark (Austrian regional subsidy programme)

Information about the regional subsidy scheme in Steiermark was collected through a telephone interview with Simone Skalicki, head of the *Umweltlandesfonds Steiermark*.

Programme description

The state of Steiermark provides direct subsidies for the installation of efficient biomass heating and solar thermal collectors. Heat pumps are not promoted by this programme.

This direct subsidy has to be seen in context with the housing programme in Steiermark: For new buildings, low-interest loans are granted. Persons below a certain income threshold can apply for loans, dependent on the size of their family. Other preconditions refer to the building: an energy audit is required and the heat demand must be low. Solar thermal technologies must be used and fossil fuel heating is only possible in exceptional cases. The loan can be increased by €7000 per installation by putting in REHC technologies (solar thermal, heat pump, and biomass) or by €3000 for connecting to district heating

Renovations fall into two categories for the purposes of the housing programme:

1 Comprehensive renovation work that includes several parts of the building that affect energy use. In this case, the owner of the building can decide

between a loan and a direct grant. Thresholds concerning income do not apply but the heat demand still has to be low.
2 Single renovation measures such as the replacement of the heating system. These simple renovations are promoted by direct grants that depend on a bank loan taken by the building owner. The building has to meet minimum insulation standards.

When only the heating system is replaced, an incentive through this housing programme is not possible in many cases. The reason may be that the building is beyond the maximum heat demand or below the minimum insulation standard or that the heating system is being replaced without taking a bank loan. For these cases, the state of Steiermark created the *Umweltlandesfonds* programme to provide direct subsidies.

The government decides annually on the budget for the programme. For 2009, a total budget of €7 million is planned for biomass (€3.3 million) and solar thermal (€3.7 million). The directive regulating the programme can also be amended annually. Political targets as well as administrative and technical aspects are considered. However, the basics of the programme have not changed much throughout the years.

The application process is kept very simple. There is very little pre-installation quality control. The exception is for biomass heating: the applicant has to perform an energy audit in order to prevent mistakes such as the installation of oversized boilers. The application is turned in after the installation, along with invoices from the installer and documentation of equipment certification. The verification of statements is 'outsourced' to the municipalities. For solar thermal, an additional subsidy from the municipality is required. This ensures that applications are not faked because the municipality provides on-site control. For biomass installations, the applicant has to provide an affirmation from his municipality.

The applications are then checked by external experts (for example, in the regional energy agencies) and are forwarded to the administration who are responsible for payment. Significant on-site controls are not performed.

The subsidy for solar thermal collectors consists of a basic sum of €300 plus €50 per square metre. The basic sum is €500 when at least 15m^2 is installed for water and space heating. The maximum amount granted is €2000.

The subsidy for biomass heating (pellet and wood chip boilers) amounts to 25 per cent of the investment costs and is limited to €1400.

Support from the housing programme and the *Umweltlandesfonds* subsidy cannot be combined. Since larger (multi-unit) building projects are better served by the housing programme, the direct subsidies primarily target owners of detached houses who cannot profit from the housing programme. Despite this narrow target group and the comparably low subsidy amounts, the number of applications handled is high. In 2008, 7000 applications were handled. In the first quarter of 2009, more than 900 applications for solar thermal collectors were registered.

Context

In 2008, 16 per cent (57.990m²) of the total solar thermal collector area in Austria was installed in Steiermark although only approximately 14 per cent of the Austrian population live there. The energy policy strategy in Steiermark is currently based upon the energy plan for 2005–2015. According to this plan, Steiermark aims to decrease the energy demand in the household and industry sector by 1 per cent annually and to increase the share of renewables in the total energy consumption from 25 per cent to 33 per cent. Energy efficiency and renewable heating uptake in the residential sector are addressed by the general building code, the housing programme and, especially for the renovation sector, the direct subsidies from the *Umweltlandesfonds*.

The improvement of installation and equipment quality is not in the focus of the programmes. The technical requirements for the programme are kept very simple and the related directives mainly rely on existing laws, codes, standards and equipment certification standards. While the housing programme covers all new buildings, the subsidy scheme targets projects where only the heating system is replaced or REHC technologies are added to the existing heating system but no further renovations are implemented to improve the buildings' energy efficiency.

The total amounts of subsidy provided by the *Umweltlandesfonds* are rather low compared to other Austrian states. However, they sufficiently reduce life cycle costs in order to reach payback times well below ten years.

Results

More than 7000 applications for biomass and solar thermal technologies were approved in 2008. The annual budgets are not sufficient to pay for all applications in most of the years and the budgets have to be extended or cases are handled under the next year's budget. This can actually be seen as an indicator of success since the subsidy amounts are relatively low.

Given the rather small target group of building owners who are not addressed by other programmes, the number of applications is impressive.

Design and implementation process

The *Umweltlandesfonds* was established in 1985. Solar thermal collectors have been subsidized since 1993 and subsidies for biomass heating began in 1998.

The directives are amended annually together with the budget. Both processes are influenced politically but input from the administrative and the technical (for example, from energy agencies) is respected. The amendments, however, are rather conservative. The maximum subsidy amounts, for example, have been stable since 2001. Other conditions for funding, such as the mandatory energy audit for biomass heating, have existed since 1998.

The directive from 2005 introduced a basic sum amount plus subsidies per square metre for solar thermal collectors and owners of multi-unit dwellings can now also apply for funding. It was decided that the largest part of the application handling (especially the technical assessment) should be outsourced to the regional energy agencies.

The continuous updating of technical requirements in the directive is not necessary since these requirements are very basic and rely on existing requirements of other laws.

The application process is kept deliberately simple. The applicant is only obliged to perform an energy audit (including a heat demand calculation) when installing biomass heating. This prevents oversized boilers from being installed. In the case of solar thermal collectors it is assumed that every installation provides advantages (financially and energetically) to the applicant. The assessment of applications is outsourced to external partners, keeping the administrative efforts at a minimum. Finally, the programme operates almost without any controls after approval. On-site inspections are not performed. Instead, the municipalities are relied upon to provide some control in the application process.

Evaluation process

The *Umweltlandesfonds* provides reports on its activities annually to the government. In addition, the *Umweltlandesfonds'* work from 2002–2007 was thoroughly assessed by the general accounting office (*Landesrechnungshof*) in 2007. The budget-handling, the design of the directives and the organization of the application handling were assessed and detailed recommendations were provided.

Challenges and barriers

Applicants contact the *Umweltlandesfonds* for the first time when the installation is complete. This simplifies the application process, however, the calculation of the necessary budget is more difficult, which is one reason for the annual shortfall of funds.

The low number of personnel and funding bottlenecks can increase the time needed for approval of applications significantly. On-site quality controls and other measures for quality improvement and control are not possible.

Programme conclusions

The housing programme in Steiermark covers primarily new buildings and comprehensive renovation projects. However, the largest potential for REHC uptake lies within existing buildings which are not renovated comprehensively and which often do not qualify for the housing programme. When biomass stoves or solar thermal collectors are installed in addition to the existing heating system or when the heating system is changed completely without taking loans, buildings do not meet the heat demand requirements of the housing programme.

For these cases, the *Umweltlandesfonds* provides direct subsidies for the installation of solar thermal collectors and biomass heating. The amount of subsidies is low compared to other Austrian states. It can be assumed that these subsidies are not the main reason why building owners decide to install REHC technologies. Persons installing REHC with the help of these subsidies are likely to be interested in renewable technologies from the beginning. The subsidy only guarantees that the payback time is reduced in order to ensure an acceptable level of REHC financial risk.

Analysis

As was found in Salzburg, subsidy programmes will often 'miss' some potential households. The *Umweltlandesfonds* targets households not covered by the housing programme. This shows the importance of a multifaceted approach. It is noteworthy that a relatively low subsidy attracts a large number of applications. The lack of on-site follow-up simplifies the application process but allows the possibility that installations are not high quality. Other factors in the programme success may be the partnerships with municipalities, stable funding and simple application process.

Key lessons learned

The *Umweltlandesfonds* regional subsidy programme suggests the following key lessons:

- The high potential in the existing building sector should not be neglected.
- This potential can be tapped with rather low costs: comparatively low subsidy amounts are enough to initiate a large number of installations.
- Programme costs can be further reduced by limiting administrative efforts.
- An obligatory energy audit before installation is an easy way to ensure pre-application quality control.
- Cooperation with municipalities and outsourcing quality control reduces administrative efforts and simplifies the application process.
- The application process is further simplified by applying a minimum of technical requirements. The programme relies on other laws and quality standards as well as the qualification of installers to maintain the installation quality.

Marktanreizprogramm (Market Incentive Programme), German Financial Incentive Programme for Renewable Heat

Information on the Marktanreizprogramm or Market Incentive Programme (MAP) is widely available in electronic and printed form. The following resources were used in preparing this summary:

- Bundesministerium für Umwelt, Naturschutz und Reaktorsicherheit, *Richtlinien zur Förderung von Maßnahmen zur Nutzung erneuerbarer Energien im Wärmemarkt*, Berlin, 20 February 2009.
- Bundesamt für Wirtschaft und Ausfuhrkontrolle, *Übersicht über die Basis, Bonus und Innovationsförderung im Marktanreizprogramm*, April 2009.
- Zentrum für Sonnenenergie und Wasserstoff-Forschung Baden-Württemberg (Dr Ole Langniß, Dr Astrid Aretz, Helmut Böhnisch, Friedhelm Steinborn), Fraunhofer Institut Systemtechnik und Innovationsforschung (Edelgard Gruber, Wilhelm Mannsbart, Dr Mario Ragwitz), *Evaluierung von Einzelmaßnahmen zur Nutzung erneuerbarer Energien (Marktanreizprogramm) im Zeitraum Januar 2002 bis August 2004*, Forschungsvorhaben im Auftrag des Bundesministeriums für Umwelt, Naturschutz und Reaktorsicherheit, Stuttgart, Karlsruhe, December 2004.
- Zentrum für Sonnenenergie und Wasserstoff-Forschung Baden-Württemberg (Dr Ole Langniß, Helmut Böhnisch, Alexander Buschmann, Dr Frank Musiol), Technologie und Förderzentrum im Kompetenzzentrum für Nachwachsende Rohstoffe (Dr Hans Hartmann, Klaus Reisinger, Alexander Höldrich, Peter Turowski), Solites Steinbeis Forschungsinstitut für solare und zukunftsfähige thermische Energiesysteme (Thomas Pauschinger), *Evaluierung von Einzelmaßnahmen zur Nutzung erneuerbarer Energien (Marktanreizprogramm) im Zeitraum Januar 2004 bis Dezember 2005*, Forschungsvorhaben im Auftrag des Bundesministeriums für Umwelt, Naturschutz und Reaktorsicherheit. Stuttgart, Straubing, October 2006.
- Zentrum für Sonnenenergie und Wasserstoff-Forschung Baden-Württemberg (Helmut Böhnisch, Tobias Kelm), *Evaluierung von Einzelmaßnahmen zur Nutzung erneuerbarer Energien (Marktanreizprogramm) im Zeitraum Januar bis Dezember 2006*, Forschungsvorhaben im Auftrag des Bundesministeriums für Umwelt, Naturschutz und Reaktorsicherheit, Stuttgart, July 2007.
- Tobias Kelm, Dr Harald Drück, Dr Ole Langniß, *Evaluierung von Einzelmaßnahmen zur Nutzung erneuerbarer Energien (Marktanreizprogramm) im Zeitraum Januar 2007 bis Dezember 2008*, *Kurzbericht für den Zeitraum Januar bis Dezember 2007*, Forschungsvorhaben im Auftrag des Bundesministeriums für Umwelt, Naturschutz und Reaktorsicherheit. Stuttgart, March 2008.

- European Solar Thermal Industry Federation, *Financial Incentives for Solar Thermal. Guidelines on Best Practice and Avoidable Problems*, Brussels, 2007.
- Sebastian Hack, *International Experiences with the Promotion of Solar Water Heaters (SWH) on Household-level*, Mexico City, October 2006.

Programme description

The *Marktanreizprogramm* or Market Incentive Programme (MAP) is the main instrument of the German federal government to stimulate the uptake of renewable heating and cooling technologies. The programme was established in 1999 and, at the time, included some support for renewable electricity. Since 2005, only renewable heating and cooling technologies are supported under the MAP. It is estimated that, since the start of the MAP, 90 per cent of the collector area installed in Germany benefited from financial support under the MAP.

The MAP has been modified ten times since its inception. The modifications changed the level of support for the different technologies and applications, as well as the requirements for eligibility under this programme. The modifications also included changes to the application procedures and processing.

In its latest edition, the MAP financially supports the installation of solar thermal, biomass heating and heat pump systems. For smaller installations, the incentives are given in form of a grant per square metre of solar thermal collector area, kilowatts of biomass heat capacity or square metre of floor area, in the case of heat pump systems. These 'basic grants' can be complemented by substantial 'bonus' payments for innovative or especially efficient systems, or the combination of different renewable energy technologies. For large installations, the MAP provides attractive loans (fixed interest rates, no pay-back in the first years, and so on).

For 2009–2012, the MAP has an annual budget of €500 million. The basic grant for solar thermal systems varies between €38.5/m^2 and €105/m^2 depending on the application, the collector area and the building age (new or existing). For the different biomass heating systems the specific basic grant varies from €27 to €36/kW for systems using wood pellets, and a flat incentive from €750 to €1125 for systems using woodchips or split log gasification. The basic grant for heat pump systems varies from €3.75 to €20/m^2 of floor area.

In 2007, the MAP incentives accounted for 8.3 per cent of the total investment in the supported installations. In comparison, in 2007 the applicable VAT rate was 19 per cent. This means that the net cash flow into public funds was positive.

Context

The MAP was introduced in 1999. The aim of the programme is to increase the use of renewable energy to protect the environment and the climate by providing incentives for renewable energy technologies.

A predecessor programme had existed since 1994, however, its financial backing was insufficient and it therefore did not have a significant impact. The first red-green (Greens/Social Democratic Party) government at the federal level took office in 1998 and implemented elements of an ecological tax reform, primarily in energy taxes. The government vowed to redistribute the additional income to foster ecological behaviour. In this context, additional financial support for renewable energies was made available. While the MAP remains the cornerstone of the federal government's support to renewable heating and cooling, it is important to note that flanking measures, such as awareness-raising campaigns, play important roles.

Results

At the end of 2007, the German Federal Ministry of the Environment announced the 1 millionth grant application in the Market Incentive Programme. Since 1999, 625,000 installations with a total investment volume of € 6.5 billion have been supported.

The small-scale installations supported with MAP grants in the rather low performing year of 2007 alone are estimated to reduce CO_2 emissions by 400,000 tonnes per year (8 million tonnes over a lifetime of 20 years). In 2008, 260,000 applications were added (compared to 155,000 in 2007) and in the first four months of 2009 the application numbers exceeded those of the same months in 2008.

Despite the relatively low funding level (approximately 8–16 per cent of the total installed cost over the years), the Market Incentive Programme has become a key driver for the adoption of renewable heating and cooling solutions in Germany. Changes in the MAP have sometimes caused dramatic changes in the market – in both directions. For example, the stepwise lowering of the specific grants in 2006 and the depletion of the MAP budget in autumn of that year contributed to the strong decrease of demand for solar thermal systems in 2007.

Design process

The MAP was designed with the weaknesses of its predecessor programme in mind, especially the very limited budget available to the previous programme: the total budget of the so called '100-million-programme' (because it provided DM 100 million, approximately €51 million) was used up after just two years.

The MAP was designed by the new red-green government that took office in 1998. Additional tax money from the increased electricity tax was to be used for the promotion of renewable energies. The budget was to be allocated each year for the following year.

The MAP is under the control of the Federal Ministry of the Environment (unlike its predecessor, which was run by the Economics Ministry) and administered by the Federal Office for Economy and Export Control (BAFA) (for the direct grant part of the programme) and the public KfW bank (for the loan part of the MAP).

In order to avoid price increases of renewable heating installations, the programme was designed to provide a financial incentive per kilowatt or square metre; rather than a fixed percentage of the overall costs. From the beginning of the programme, technical quality requirements were established and have been adapted and tightened over the years.

Because financial incentives can only provide one piece of the market stimulation puzzle, the MAP was complemented with awareness raising campaigns (*Solar – na klar!* 1999–2002, *Solarwärme Plus* since 2002, *Wärme von der Sonne* since 2005 and *Woche der Sonne* since 2007). Furthermore, to promote the technological research and development, programmes such as *Solarthermie2000* and *Solarthermie2000plus* provided financial support for pilot and demonstration projects.

Implementation process

Due to the rapidly growing success of MAP, the efficient processing of grant applications became an important task for BAFA. Today, the agency employs more than 100 people to administer the more than 250,000 applications received per year. Several times the process was simplified in order to reduce administrative overhead and waiting times for the applicants.

Beginning in 2007, the grant application process happened in two stages: the application was first filed with BAFA, whose staff reviewed the applications and authorized grants. Once the installation of the renewable heating system was completed, the investor informed the BAFA and claimed the allotted grant. The BAFA then executed the payment.

Until 2007, the grant application process required only one stage: after the installation is completed, the application was sent to BAFA, which approved the grant and executed the payment. In order to provide investor security, BAFA publishes a *Förderampel,* or grant signal, which shows how much of the available budget has been used so far and whether it is still safe to install a system in order to then receive the grant.

Even before the implementation of the one-stage application process, the applications for solar thermal systems had been simplified. An important decision in this direction was to calculate the grant based the 'rounded up' square metres of collector area. (For example, an application for 5.1m^2 of collector area would be awarded a grant based on 6m^2 of collector area.) In the beginning the grant had been calculated on the exact collector area. However, in many cases the actual installed system deviated slightly from the one applied for, which meant that BAFA had to re-evaluate the approved grant. By basing the grant on the 'rounded up' square metre instead, a recalculation can often be avoided.

Twice, in 2001 and 2006, the specific MAP subsidies were reduced at short notice because application rates indicated an early depletion of the available budget. In both cases, the solar thermal market showed a strong decline in the following year. While other factors clearly contributed (such as the introduction of the euro currency in 2002, the international insecurity after the 2001 terrorist attacks in the US and the increased VAT rate in 2007), the reduction of the MAP grant was seen as one of the reasons for the decrease.

Experience with MAP and with financial incentive schemes in other countries showed that the discussion of changes to a grant scheme has strong and negative effects on market stability. In cases where a decrease of the grants is discussed, consumers rush to still receive the (higher) grants, leading to a temporary overheating of the market and then a phase of sharp decline. When an increase in grants is discussed, consumers typically postpone their investment decision until after the new regulations come into effect. This can bring the market to an almost complete halt until after the new grants are in place. The German government now announces changes with lead times of just a few days. In 2006, when the MAP budget was depleted by autumn, the government took the unusual step of announcing that in the first months of 2007, applications could be filed for those systems installed already in 2006, when no budget was available. This helped prevent a drastic downturn otherwise expected at the end of the year.

The strong effect of changes to the MAP became also very visible in 2005 when the specific grant for solar thermal collector area was divided into a lower one for domestic hot water-only systems and a higher one for combisystems, which also provide part of the space heating demand. A steep increase in installations of combisystems was evident after this incentive change, even though the additional incentive amount was very small compared to the additional cost of a solar combisystem compared with a solar domestic hot water-only system.

In 2009 the Renewable Heating Law came into effect. The law requires the use of renewable energy to cover a certain minimum percentage of the total energy demand of a new building. The specific percentage depends on the technologies used. The new MAP regulations significantly lower incentives in new buildings and only provide incentives when the installation provides a higher percentage of the energy demand than what is legally required.

Evaluation process

The MAP has been regularly evaluated by several independent scientific institutes (ZSW, Solites, ITW and others). Detailed evaluation reports exist for 2002–2004, 2004–2005 and 2006. A short version for the first year of the two-year period 2007–2008 was published in 2008.

The evaluations aim at answering the following questions:

- How much is the MAP responsible for the market penetration of the supported technologies?
- How high is the financial incentive relative to the total cost of the installation?
- Have cost reductions been noticed and how much are they due to the support programme?
- What is the contribution of the MAP towards greenhouse gas emission reductions?

The studies are based on the available data from BAFA and KfW and have frequently been complemented by interviews and questionnaires of market participants (planners, installers, manufacturers, industry associations and end users).

The evaluation reports not only study the workings of the previous period but also give recommendations for the further improvement of the programme. Many of the recommendations, including the one to move towards a one-step application process, were later taken up by the government.

Challenges and barriers

The main challenge for the MAP has always been the funding situation. Unlike the highly successful Renewable Energy Law for the electricity sector, which provides for feed-in tariffs paid by the grid operator, the MAP is financed from the public budget and therefore needs annual approval. This has made the programme unstable and created uncertainty in the market, thus disrupting the otherwise strong development of the market for renewable heat and cooling. The depletion of the annual budget contributed to the significant decreases of the solar thermal market in 2002 and 2007. The Renewable Heating Law, which was approved in 2008, was intended to provide more stability at lower cost to the public budget.

Another constant challenge is the adaption of the technical/quality requirements to the development of the technology and markets. While very strict requirements would ensure the highest quality, they could become a barrier if only a few products were eligible or if the administrative burden was too high. On the hardware side, the requirements for solar thermal have been increased from just a test of the collector according to the applicable European Standard (EN 12975) and a collector yield of 350kWh/m^2 per year to a full Solar Keymark certificate on the collector and the fulfillment of the RAL Gütesiegel (2004 edition), including a minimum collector yield of 525kWh/m^2 per year. However, the quality of other components is not adequately addressed and there is no check of the actual installed system to ensure functionality.

Programme conclusions

The MAP is the cornerstone of the German federal government's measures to increase the use of renewables in heating and cooling. After almost ten years of operation (with approximately one change to the MAP per year), its administration runs very smoothly and waiting times between applications and approval/payments are typically only a few weeks. The programme is well-known to all professionals, who typically advise their customers on the availability of the MAP financial incentives. The market reaction to changes in the MAP indicates its high impact on the market.

Analysis

The MAP programme is well-known as a comprehensive, effective REHC programme, the success of which is reflected in Germany's high REHC installation rates. The programme is notable for a number of reasons:

- The commitment is long-term.
- The programme has been regularly evaluated and modified based on those evaluations.
- The programme is multifaceted, combining incentives with education campaigns and pilot and demonstration projects. A 'stick' has now been added to the mix with the 2009 regulation requiring renewable heat in some new buildings.

Key lessons learned

The following conclusions can be drawn concerning the MAP:

- A well-managed grant programme such as the MAP can have a very significant effect on the uptake of renewable heating and cooling solutions.
- The fact that the programme has been running for many years has helped create awareness of and confidence in the MAP.
- The effect of the MAP by far surpasses the specific financial incentive provided: the fact that the government offers support for the installation of renewable heating/cooling installations is a strong psychological/marketing tool.
- Changes to a long-running financial incentive scheme are absolutely necessary, for example, to update the technical requirements. However, they must be made only after close consultation with the market actors and announced at the last moment in order not to disrupt the market.
- A financial incentive programme can be successfully used to promote certain applications or technologies (such as those with a higher-than-usual annual yield) and they can complement a renewables obligations in the building sector.
- A financial incentive scheme does not make other measures irrelevant (such as awareness-raising, education/training of professionals, support for research and development).

The Solar Keymark Certification Scheme for Solar Thermal Products

Information on Solar Keymark was provided by Uwe Trenkner of the European Solar Thermal Industry Federation, who has been directly involved in the development of the Solar Keymark certification scheme since 2003.

Programme description

The Solar Keymark is not an independent support scheme, rather it is a tool that allows authorities managing solar thermal incentive programmes (financial or regulatory) to assure a high quality of the products under the support scheme. The Solar Keymark programme has helped to establish common requirements for most European solar thermal support schemes and to remove barriers to trade within Europe.

The Solar Keymark is a certification mark owned by the European Committee for Standardization (CEN). It certifies conformity of a product with the applicable European Standards: EN 12975 for solar thermal collectors or EN 12976 for factory-made solar thermal systems.[8] The Solar Keymark is the solar thermal specific edition of the general Keymark certification scheme, which is available for a wide variety of products and services.

The Solar Keymark has the following main requirements:

- The test sample of the collector or system is chosen out of the production or warehouse by an independent inspector (typically from an accredited testing laboratory).
- The test, according to the applicable European Standard (EN 12975 or EN 12976), is carried out by an independent, accredited testing laboratory.
- The manufacturer must have implemented a factory quality management system similar to the ISO 9000 series of standards.
- A sample of the product, again selected out of the production or warehouse by an independent inspector, is visually inspected every other year to ensure that the product is still the same that was originally certified.

Currently, there are five different certification bodies authorized by CEN to issue the Solar Keymark. Other certifiers can apply to CEN to also become authorized to issue this specific Keymark. Each of them has cooperative agreements with one or more accredited testing laboratories, which carry out the actual test and often also the inspection of the factory.

The Solar Keymark is used in many solar thermal support schemes in Europe as a criterion that allows a product to gain eligibility to the support scheme. Some countries have brought their own national scheme in line with the Solar Keymark (for example, Sweden, Portugal), while others support products that are either Solar Keymark certified or have received a similar national certificate (for example, France, the UK). In Germany, the Solar Keymark on the collector is a necessary prerequisite to receiving grants under the Marktanreizprogramm (Market Incentive Programme or MAP).

The Solar Keymark scheme rules, as well as the underlying standards, are approved and owned by CEN and the CEN Certification Board. However, any changes are typically developed by the Solar Keymark Network, which groups together the involved certification bodies, testing institutes and representatives of the solar thermal industry in Europe.

Context

In the 1990s, many European countries had established financial support programmes for solar thermal systems, significantly increasing the market. In the absence of European standards, each country implemented its own technical requirements for the supported products. This posed a serious barrier for companies from other countries because they had to fulfil various additional tests for each country in which they were attempting to sell products. The agreement on a European standard in the late 1990s reduced the problem but many countries kept their own requirements or demanded that foreign-made products be (re) tested in their own country. Without easy access to local support programmes, foreign manufacturers found it difficult to compete with domestic manufacturers.

In 2000, the European Commission co-funded a project under their ALTENER Programme to develop a European certification scheme for solar thermal products, based on the CEN Keymark. In the framework of this project, the solar thermal industry, together with test institutes from several European countries, developed the specific Solar Keymark scheme rules, which were officially agreed by the CEN Certification Board in January 2003. A few months later, DIN CERTCO became the first certification body empowered to issue the Solar Keymark. In October 2003, the first Solar Keymark was issued for a solar thermal collector. In February 2005, a factory-made system received the first Solar Keymark.

In the first years of existence for the Solar Keymark, a chicken-egg problem existed: most solar thermal companies did not see a good reason to pay for the Keymark certification as there were almost no support programmes which linked eligibility to this certificate. Public authorities in charge of solar thermal support programmes, on the other hand, did not take the Solar Keymark seriously as there were almost no Solar Keymark certified products out in the market. Only in 2006 did the Solar Keymark really take off, and now there are more than 700 Solar Keymark certified products available on the market.

Results

The Solar Keymark has become an important tool to open national markets to quality products from other countries. By making it easier for Solar Keymark products to be eligible for national support schemes, governments help establish a high product quality and increase choices for the end-consumers.

Today, there are more than 700 Solar Keymark products in the market, which is estimated to be more than half of the products traded across borders in Europe. The broad acceptance of the Solar Keymark as an eligibility criterion in support programmes has significantly reduced testing costs to the manufacturers or importers and sped-up the introduction of new products in many markets.

Design process

The Solar Keymark was developed largely by the solar thermal industry in order to overcome the trade barriers resulting from different requirements in different support schemes. The two then existing solar thermal associations at European level, ESIF and ASTIF (which merged into ESTIF in 2002), were both involved in the ALTENER Solar Keymark project that worked out the specific Solar Keymark scheme rules. The testing and research institutes, which would in the short-run lose some testing business due to the overall reduction of tests needed in Europe, fully supported the introduction of the Solar Keymark, recognizing that such a common certification scheme would eventually be beneficial for the overall market development.

The general requirements of the Solar Keymark certification scheme are fixed through the Keymark rules established by CEN. The focus was on working out the specifics of solar thermal products. Among the many issues that needed to be addressed were:

- The general Keymark rules foresaw product tests every other year, but there was general agreement that this would raise the cost of the Solar Keymark to a level that would be unacceptable to most of the industry. In close cooperation with the CEN Certification Board, it was agreed to instead establish a 'visual inspection', which would sufficiently ensure that the product is still the same as the one initially certified.
- Several of the test institutes are or were public or semi-public institutes. As the Solar Keymark would do away with local testing requirements, it was feared by some test institutes that the subsidized institutes could offer the lowest testing cost and thus gain most of the market. A transition period was agreed upon after which no public money was allowed to subsidize the testing business.

After several rounds of discussion between the stakeholders, a final version of the scheme rules were proposed to the CEN Certification body, who approved them in early 2003.

Implementation process

The implementation followed the process outlined in the (Solar) Keymark scheme rules. Certification bodies everywhere in Europe could apply to CEN for authorization to issue the Solar Keymark. The first authorized certification body was the German DIN CERTCO, which is the certifier with the highest number of issued Solar Keymarks. (This is not only due to its early interest in the certification but also to the fact that more than 50 per cent of the solar thermal market in Europe and its manufacturing capacity are concentrated in the German-speaking countries of Germany and Austria.) SP (Sweden), CERTIF (Portugal), ELOT (Greece) and ICIM (Italy) are also authorized certification bodies. Other certification bodies have voiced interest in participating and are considering application for authorization.

There are currently 18 accredited test laboratories that carry out tests according to EN 12975 (collectors); and 11 accredited test laboratories that test according to EN 12976 (factory-made systems). This allows manufacturers or importers to chose a test institute of their liking, perhaps because they have prior experience in working with them, or because they offer faster service.

In 2006, the European Communities, under their Intelligent Energy-Europe Programme, have co-funded a follow-up project (SolarKeymark-II, 2006–2007), to evaluate, improve and promote the Solar Keymark. Within this project, the stakeholders of the Solar Keymark certification process (certification bodies, testing laboratories and the solar thermal industry) set up a Solar Keymark Network, which has met once or twice per year to discuss the further development of this certification scheme and to agree on a common application of the scheme rules as well as the underlying standards. The network aims at having a harmonized application of standards and scheme rules and to thus ensure an equal, high quality of the whole certification process everywhere in Europe.

Evaluation process

An evaluation of the Solar Keymark was carried out in the SolarKeymark-II project. It analysed several aspects of the programme, such as the acceptance of the Solar Keymark by both the industry and the public authorities, and current shortcomings of the testing and certification. It also provided recommendations on further improvements.

A formal evaluation report is not foreseen for the future, but the Solar Keymark Network allows any of the stakeholders in the Solar Keymark certification process to point out potential problems with the current practice and to propose how solve these issues.

Challenges and barriers

The current main challenge is the very inflexible system certification: if the collector area or the tank volume of a solar thermal system increases or decreases, a new certification is necessary. As many manufacturers offer a wide variety of combinations of different numbers of the same collector and same tanks but different volumes, this inflexibility creates high costs in terms of (waiting) time and money paid for each individual test and certification.

A new proposal for a flexible system certification allowing, within certain limits, an upscaling or downscaling of results for the same products but different collector area or tank volume is currently being developed and discussed among the experts. An agreement on a simple but still reliable calculation method is expected to lead to a much increased demand for system Keymarks. Several EU member states, including France and Portugal, have already shown high interest in making the Keymark for the whole system a requirement in their support programmes.

Furthermore, it remains a challenge to extend and update the standards and scheme rules to the latest technological and market developments. For example, today there is no European Standard covering air collectors, and therefore no Solar Keymark can be issued for this technology.

Another challenge is the adoption of the Solar Keymark by public authorities developing solar thermal support schemes. In several national support schemes, additional requirements remain. (For example, for roof-integrated collectors, the UK still requires a test for the weather-proofing of the collector and in Germany the government requires the fulfilment of certain additional parameters set by its 'Blue Angel' eco-label.) Even where the support programme fully accepts Solar Keymarked products as eligible, other market actors may ask for other or additional certificates. (As an example, in France, housing insurance is offered by most insurers only if the solar thermal product has the French CSTBat certificate.)

Programme conclusions

The Solar Keymark has been highly successful in harmonizing technical requirements in solar thermal support schemes throughout Europe. The certification scheme has allowed manufacturers to relatively quickly and easily get their products into national support schemes and reduced overall testing and certification costs and waiting times. This has lowered the barrier to market entry for many companies and has led to more quality products being available in many national markets.

Analysis

The Solar Keymark programme is an interesting example of multi-country cooperation to solve a barrier to solar thermal installations. Although the market impact of this programme took some time to develop, the Solar Keymark now appears to be an important tool facilitating national and regional promotion programmes. Key to the programme's success appears to be the involvement of and cooperation with a wide variety of organizations.

Key lessons learned

The following conclusions can be drawn concerning the Solar Keymark certification scheme:

- A key factor for the success of the Solar Keymark was the 'early-adopters' from the industry and the national governments who realized the potential advantage of having such a common certification scheme.
- The Solar Keymark scheme rules, as well as the underlying EN standards, need frequent revision to take into account new technical and market developments.
- The general high regard for the Solar Keymark as a certificate which identifies high quality products can only be maintained if all stakeholders work together to keep up the high standards of the certification process.
- The Solar Keymark has helped ensure the high quality of the products installed in many countries in Europe. However, it cannot ensure the correct planning and installation of the certified products and thus the correct working of the overall installation.

Barcelona Solar Thermal Ordinance, Spain

Information about the Barcelona Solar Thermal Ordinance programme was collected through a telephone interview with Dr Josep Puig, an engineer and the Eurosolar vice-president, who was on the city council during the development of the ordinance. Information was also obtained from the Barcelona Energy Agency website.[9]

Programme description

The Barcelona Solar Thermal Ordinance, implemented by the municipality of Barcelona, Spain, requires that new buildings or buildings undergoing major renovations have a solar domestic hot water system that meets at least 60 per cent of hot water demand. Other forms of solar heating are not included in the ordinance. In the first version of the ordinance, in place from August 2000–2006, buildings with a hot water demand of less than 292 MJ/day were exempt from the ordinance. This effectively exempted most single-family homes. However, in 2006, the revised ordinance removed this minimum restriction and therefore the ordinance now applies to all new and renovated buildings. Other new aspects of the revised ordinance include improved maintenance and architectural requirements and a new regulation that pools be heated 100 per cent by solar heating. Best available solar thermal technology must be used and compliance must be verified with measured data. Violations of the ordinance result in fines of €6000–60,000.[10]

Some buildings are exempt from this regulation if they cannot reasonably meet 60 per cent of their solar demand through solar water heaters (for example, buildings with no significant sun exposure). In a few cases, the requirement can be met by installing other renewable technologies such as biomass, cogeneration and or waste heat recovery. However, solar hot water collectors seem to be the most appropriate technology for Barcelona, given its Mediterranean climate, and therefore the ordinance has not resulted in a significant number of installations of any other renewable heating technology.

Guidance is available through the Barcelona Energy Agency at all stages of the process. A solar thermal guide is available to the public, as are training courses for professionals and assistance in dimensioning and implementation,[11] much of which is available online. Consultation is available, both from the agency staff and from specialists hired by the agency to solve more difficult problems. The exact cost of the ordinance is difficult to isolate from the Energy Agency budget, which is completely provided by the municipality.

Context

The main sources of energy in Barcelona are nuclear-generated electricity and fossil fuels. During the 1990s, there was growing interest in renewable energies in Barcelona. The Action Plan (APE) was developed to increase the share of renewables in the energy mix and reduce emissions. There was also a desire to make renewable energy visible within the city. Solar thermal collectors benefit from the significant solar resource available in Barcelona, and therefore presented a good opportunity to reduce natural gas usage. Despite this, total installed solar water heating capacity installed in the city prior to the ordinance was less than $1000m^2$. The heat target specified in the APE is $96,300m^2$ of thermal solar panels with a heat generation capacity of 280,000GJ/year; part of a plan to provide 679,800GJ/year from renewable sources or 1.1 per cent of the city's total consumption by 2010.[12]

The ordinance was particularly timely considering the construction boom that was occurring in Spain during that period, although the Barcelona market had a less significant increase in construction than other Spanish cities. Also, the city had a significant proportion of multi-family housing. Solar hot water systems can be more cost-effective on a multi-family house because capital costs may be shared.

Results

The ordinance has been very successful in Barcelona. From 2002–2006, $40,095m^2$ of solar hot water collector capacity was authorized and, once installed, resulted in a savings of 32,076MWh per year and an associated 5640 tonnes of CO_2 per year.

An overall objective of 100,000m² of solar thermal collector surface area was set for 2010,[13] of which 40 per cent had been achieved by 2006. As it is seen in Exhibit 21, most of the growth in solar thermal installation has been in the residential sector.

Initially, there was a lot of pessimism about the programme: neither the public nor government officials believed it would work. However, the ordinance has been very successful and has been a model for more than 60 other municipalities in Spain. In 2006, the Spanish government developed a national building code that has a requirement similar to the Barcelona ordinance. The main difference in the requirements is the solar fraction requirement (percentage of hot water demand that must be met with the solar collector system), since the solar resource varies throughout Spain.

Design and implementation process

In 1997, the city council of Barcelona began discussing the idea of a solar ordinance, similar to that being proposed in Berlin, Germany. With a recent municipal election and an increased interest in sustainability, a new political identity was created to develop renewable energy projects, called the Sustainable City Councillor. Dr Puig, who filled this role, asked why the significant solar resource in Barcelona was not being used to produce energy locally and encouraged the idea of a solar ordinance.

Non-governmental organizations and solar companies were involved in the development of the ordinance, in collaboration with municipal politicians. The

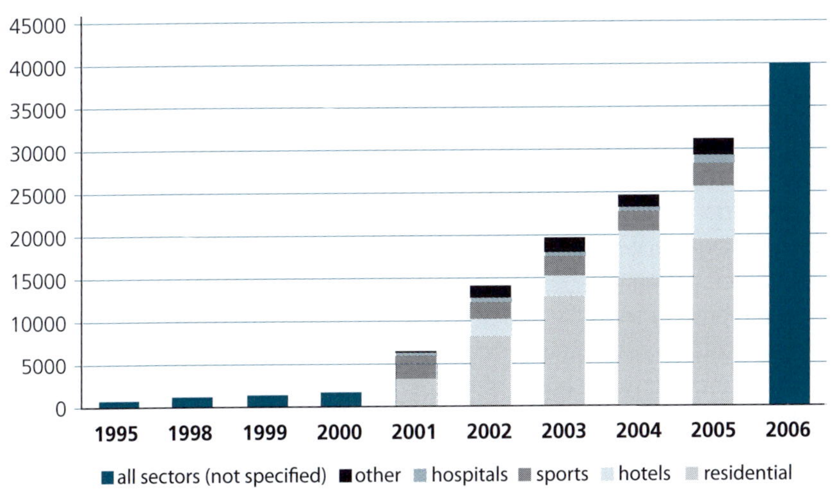

Exhibit 21 Barcelona solar thermal hot water installations (in m²)[14]

regulation was tested for one year, and then the ordinance was formally approved in July 1999. It came into effect for all requested buildings permits as of August 2000.

The Barcelona Energy Agency, originally called the BarnaGEL, became the administrator for the solar ordinance programme. The agency was started in 1995 to promote renewable energy in the metropolitan area of Barcelona. It was a project sponsored by the EU to encourage cooperation between local politicians, organizations and the public to develop energy projects. However, in 2000, funding from the EU was replaced by support from the municipality through the Barcelona Energy Agency, giving the city complete control in decisions made regarding the ordinance. The staff is made up of approximately 20 people. It is difficult to separate the costs of the agency from the costs specific to the Solar Thermal Ordinance.

One concern was that the ordinance would create a large new market which would overwhelm local solar collector manufacturers. However, since manufacturers in neighbouring European countries were easily accessible in the open market, this issue was quickly laid to rest. In the end, the Spanish market was able to handle the increase in demand internally.

The application process is not time-consuming for either the administrator or the applicant. When the applicant submits a proposal to the city in order to receive a building permit, the building design must include a solar installation. The solar installation is reviewed by the Barcelona Energy Agency. In the case of missing or erroneous solar installation plans, or of uncertified equipment, the agency works with the applicant to develop a better solar installation that meets the ordinance specifications. An inspection team is occasionally sent to the building site to ensure that the ordinance is being respected and that certified equipment is being properly installed. The agency may subcontract professionals to deal with a more complex problem. Verification is done by a certified installer. Random inspections are occasionally done by the municipality.[15]

Evaluation

The Barcelona Energy Agency issues an annual evaluation of the programme. The analysis is done by recording the number of square metres of solar hot water heating systems for which approval has been granted that year. These are compared to the targets set out by the APE. It has been suggested that the evaluation could be more thorough and take into account how many buildings have actually been built with solar systems rather than just tally the approved permits. This would allow for a more accurate count of solar systems installed in the city.

Qualitative analysis is regularly done by the municipality through general surveys conducted on the street regarding the presence of solar in the city. The response has been very positive and citizens seem to be pleased with the increasing solar capacity visible in the city.

Challenges and barriers

The initial barrier for the adoption of solar thermal in Barcelona was the capital and installation costs of solar systems. However, studies indicated that solar domestic hot water systems generally cost less than 1 per cent of the total project cost. Also, housing in Barcelona is predominantly in multi-family units rather than single-family homes, and this means that costs are shared among several homeowners. This justified the ordinance and removed this excuse for non-compliance.

Product and installation quality issues have become apparent. Since the ordinance requires that only 60 per cent of hot water demand be met with solar thermal systems, an auxiliary heater is always included in the design. Therefore, in the first years of the programme, people were often not aware that their solar water heater system was not functioning properly because they continued to have hot water available in their homes. This has been solved by improved supervision of design (both at permit stage and through site inspections by the municipality) and a requirement that installed products meet Solar Keymark standards.

Although the ordinance was deemed to be quite successful, it was felt that the regulation had a limited impact on total energy use in the municipality. Hot water heating only represents a fraction of total building energy use and Barcelona has had a limited number of buildings built or renovated since 2000. Therefore, the ordinance applies to only a small proportion of the buildings consuming energy in the city.

Programme conclusions

The Barcelona ordinance is an excellent example of a stick-based programme designed to promote the use of a suitable renewable resource in a given municipality. It has been very successful and is consistently measured against targets set in the original APE. The ordinance came at the right time and capitalized on a period of increased building construction. It was supported by mature technology, public interest and political commitment. It is a good example for other municipalities on how to capitalize on renewable resources and economic conditions in their area.

Specific features that might be useful to other regions/countries are:

- Start at the municipal level in the development of regulations.
- Use a small, local energy agency to undertake inspection, monitoring and dissemination of best practices to installers and customers.
- Regulate the quality of equipment to be installed.
- Involve the public in appreciating the renewable energy in the city. Include questions about solar collector appearance in general on-street surveys.

- Get committed politicians to participate.

Analysis

The Barcelona Solar Ordinance has been effective not only in Barcelona but also in inspiring additional solar ordinances in Spain and elsewhere. The programme appears to have developed effective solutions to application processing and providing assistance to both building owners and professionals. Other success factors appear to be pre-testing, a method of enforcement and a mature solar sales and installation market. Additional formal evaluation might improve the programme, although it appears that informal feedback has been used to correct problems as they are identified.

Some ordinance or regulation programmes have not been as successful. For example, the Executive Order Solar Heating Obligations in New Buildings Outside the District Heating Areas in Denmark is a regulation that has been in force since 2001. It produced a 5 per cent annual increase in solar thermal heating however this was a modest moderate growth rate when compared to that in neighbouring countries.[16] It is difficult to know whether this was a failing of the ordinance or whether programmes in neighbouring countries were simply more effective in promoting solar thermal heating.

Key lessons learned

The Barcelona Solar Thermal Ordinance provides the following lessons:

- Quality standards are important to ensure that the solar system is effective in generating thermal energy. Certification of equipment and supervision of design and installation are important in order to demonstrate reliability of the system and to achieve real savings.
- Three key factors were identified in the success of this ordinance:
  - political will and investment of governing bodies;
  - technical ability available to design, manufacture and install good quality products;
  - involvement and interest of the public.
- A more structured evaluation plan could give a more accurate analysis of the amount of energy being saved through this ordinance. The building process should be monitored and the evaluation should give information about how much solar collector area has been installed, not only how much solar collector area has been approved through the building permit.

France's Direct Tax Credit

Information about the French tax credit programme was collected through a telephone interview with Frédéric Tuillé at Observ'ER (the Renewable Energy

Observatory), which publishes nationally and internationally recognized studies regarding energy, the environment and development. Evaluation reports from the REFUND+[17] website were also used.

Programme description

The French tax credit programme was implemented as a direct fiscal measure for the support of renewable heating and other renewable energy investments. The tax credit applies to household solar domestic hot water and other solar heating, biomass heating, and heat pumps. The initial credit, implemented from 2001–2004, was for 15 per cent of the cost of the equipment. The second time period of the credit (2005–2009) allowed a 40 per cent credit, which was then amended to 50 per cent (as of January 2006). For 2009, the credit for biomass heating has been reduced to 40 per cent because the market has been adequately developed; it now requires a smaller incentive to maintain consumer interest.

For the 2001–2004 programme, cap values for investment[18] had been set as:

- €3077 per single person.
- €6,154 per couple.
- Plus€308 per supplementary person, €384 per second child and €462 per third child.

For the second period, 2005–2009, cap values were adjusted to:

- €8000 per single person.
- €16,000 per married couple.
- Plus €400 per supplementary person, €500 per second child and €600 per third child.

Restrictions included:

- The equipment must be installed at the primary residence.
- The equipment must be installed by a professional (roughly defined as someone who is able to provide an invoice).
- As of 2008, renters were allowed to apply for the credit if they purchased the equipment. (Prior to 2008, only homeowners were eligible.)

Guidance was also provided, in the form of radio and TV advertisements for the public and specific information for industry members. It was recognized that all market players should be involved in the tax credit programme in order to promote and implement it as much as possible. Training was also provided to installers through renewable energy sector associations, building workers associations and industrial associations. The French Environment and Energy Management Agency (ADEME) also offers some training, but is limited by its staffing capacity and location.

In 2004, the total costs of the tax credit were announced by the French Inland Revenue Agency as €4.5 million for solar thermal, €27.5 million for GSHP, €37.4 million for wood energy for a total of €69.4 million. Approximately 70 per cent of the cost of the tax credit was recuperated in the VAT collected on the purchases of the renewable heating systems.[19]

The costs for other years are shown in Exhibit 22 below:

Context

France has limited fossil fuel reserves and sources of energy. National energy policy has traditionally been focused on demand management and energy efficiency rather than renewable energy. The national policy of renewable energy that does exist is more geared to electricity and biofuels. However, following EU targets and growing national interest, the tax credit was implemented as a result of the Finance Act 2001.

Results

In 2004, during the first period, approximately 35 per cent of sales of solar thermal equipment took advantage of the tax credit. This first period also coincided with Plan Soleil, which was an initiative that offered guidance and direct subsidies from ADEME equal to the tax credit for solar thermal systems. Most purchasers did not take advantage of the tax credit because they did not know about it. Approximately 60 per cent of wood/biomass heating purchasers used the tax credit, which indicates that the subsidy may have been an important driver in the growth of the residential biomass market. Finally, for heat pumps, the proportion of buyers who took advantage of the tax credits was 90 per cent, indicating that the tax credit was definitely a driver in the growth of the residential heat pump market.

Several regions offered additional subsidies, which meant that for certain renewable energy technologies, the regional subsidy plus the federal tax credit accounted for 80 per cent of the capital cost. In order to avoid offering this

	Total amount for 2004	Total amount for 2005	Total amount for 2006
Solar thermal	€4,500,000	€25,100,000	€93,000,000
GSHP	€27,500,000	€103,000,000	€324,000,000
Wood energy	€37,400,000	€117,000,000	€269,000,000
Total	**€69,400,000**	**€245,100,000**	**€686,000,000**

Exhibit 22 Programme costs for French direct tax credit[20]

double subsidy to consumers during the second time period (2005–2009), more emphasis was put on the tax credit as the main incentive for renewable heating in households. Regions are shifting their subsidies towards energy efficiency measures, collective installations and loan programmes and have abandoned direct subsidies for renewable heating in households.

During the second period to date, approximately 95 per cent of purchasers of solar thermal equipment and heat pumps took advantage of the tax credit. During interviews, programme participants indicated that the tax credit was the most significant factor in their decision to purchase equipment, followed by the increasing cost of fossil fuels and other federal and regional renewable energy subsidies and campaigns.

The increase in solar heating capacity is shown in Exhibit 23. The income tax credit implementation and programme changes are also shown. Also note that a solar thermal national programme, Plan Soleil, was implemented in 2000. It can be seen that solar thermal capacity has greatly increased since the tax credit was changed to 40 per cent in the second phase (2005–2009) of the programme. This is probably also due to the increased public awareness of solar due to the implementation of Plan Soleil, although the tax credit may have been the main driver in the market growth.

The increase in biomass heating as shown in Exhibit 24 is also more pronounced since the change of the tax credit rate to 40 per cent. A national biomass programme, Bois-Energie, was also started in 2000 and most likely contributed to the great increase in capacity in 2005 and 2006, although the tax credit may have been the main driver in the market growth.

In Exhibit 25, heat pump sales are shown. It can be seen that the heat pump market has had more steady market growth despite the tax credit rate increase. This is probably because the market responded well to the credit from the beginning. (The tax credit was requested by 90 per cent of purchasers in 2004, much higher than for solar or biomass.)

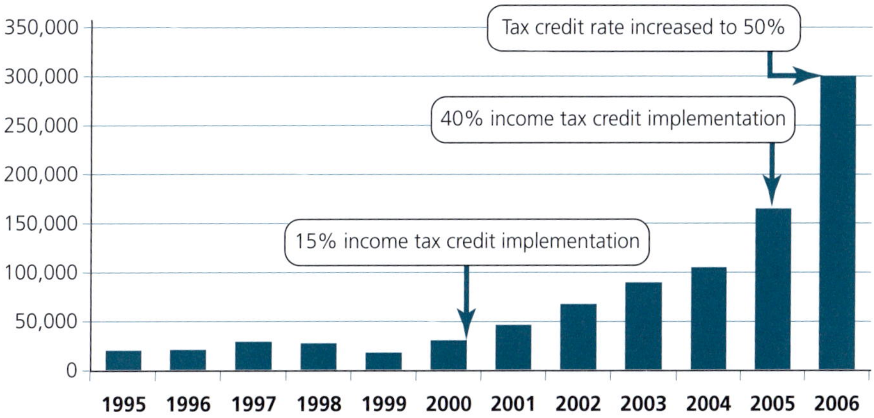

Exhibit 23 Evolution of the French solar thermal energy market (in m²)[21]

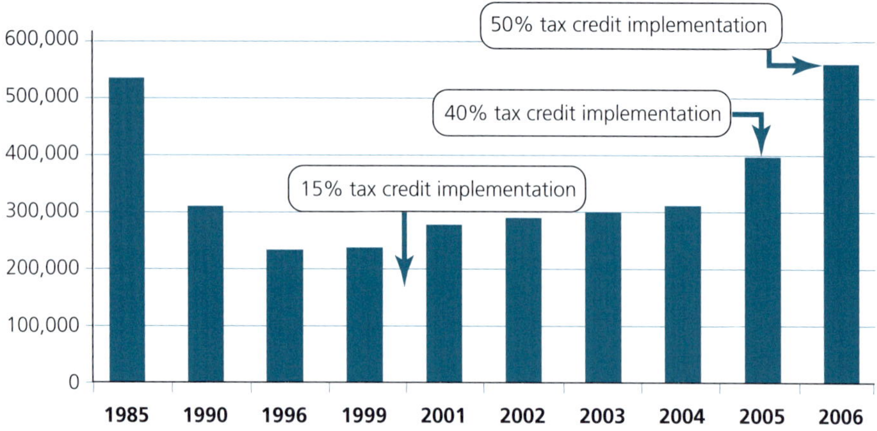

Exhibit 24 Comparison of the total biomass heating market for 2005 with the operations helped by tax credit[22]

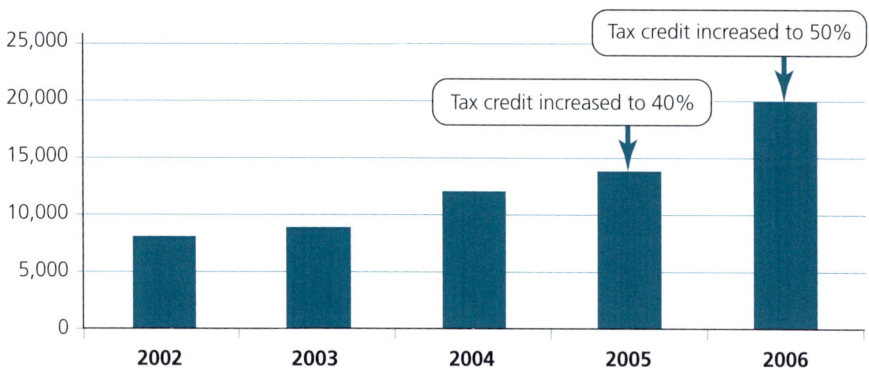

Exhibit 25 Comparison of the total heat pump market for 2005 with the operations helped by tax credit[23]

The tax credit programme has been considered successful qualitatively because the market for each of the three technologies was developed and equipment quality was increased. One of the main objectives was to strengthen the industry structures for each technology, particularly for biomass, for which the distribution chain for fuel is important but difficult to develop.

Design and implementation process

In the mid-1990s, France began planning income tax credits as part of developing a national energy policy. The overall goals were: the liberalization of the electricity and gas market; modifying underlying trends in urban transport planning sectors to control fuel consumption; preparing to replace fossil fuel electricity generation plants while leaving the nuclear option open; and increasing the fairness of energy tax policies. Finance Act 2001 initiated this first tax credit for renewable energy.

The programme was initially designed based on German and other EU renewable energy programmes. A tax credit design was chosen because of the minimal administration costs and overhead required to implement the programme. Past experience with subsidies in France indicated that the administrative costs were prohibitive and people were discouraged from applying if the subsidy application approval process was complicated. Instead, as a tax credit on an annual tax return, a household could claim up to 15 per cent of the equipment cost on their tax return with minimal administration and organizational costs.

A campaign for the public and all market actors in the three key renewable energy sectors was developed. A public TV and radio campaign made the public aware of the tax credit. Specific information was provided to all market actors. The important role of installers was recognized, since they provide an important link between the public and industry. Training and information was therefore provided for them to promote the tax credit to their customers, which took two to three years to fully implement.

The first tax period, 2001–2004, acted as an initial test of the tax credit. For the 2005–2009 period, several changes were made, including increasing the value of the subsidy and adding more specifications on quality products. In the next period, recommendations include modifying the subsidies for each technology, depending on:

- Market price: determine how much incentive is needed to be attractive to potential buyers.
- Efficiency: in particular for wood burning equipment, give higher incentives to more efficiency equipment.

Evaluation process

Several evaluations have been conducted by Observ'ER and project partners, supported by the EU Intelligent Energy initiative. Evaluation has included assessments from the French Inland Revenue Agency and interviews with participants regarding their satisfaction with the programme and adaptations they have made to their energy behaviour. The programme has been directly compared to similar tax credits in several other European countries.

Challenges and barriers

The initial barrier for these technologies was primarily cost. This barrier was dealt with directly by implementing subsidies and indirectly by higher fossil fuel prices.

The second barrier was that subsidies were expensive to governments due to the high administrative costs. The high administrative overhead was also a hassle for potential applicants. Both of these issues were solved by implementing a direct fiscal measure rather than a subsidy programme.

One challenge that was noted was that customers were given different information by installers, manufacturers and tax officials. In some cases, customers were not clear on what equipment was eligible for incentives. For heat pumps, for example, only the outdoor equipment is eligible for the credit. It has been suggested that it would be better to have a centralized agency to provide information to all market players and customers.

Another challenge was the market distortion that occurred as a result of the programme. Since installation costs were not eligible for the credit, certain installers provided 'free' installation for overpriced equipment, presumably in order to take advantage of the tax credit. This happened in the case of heat pumps, but did not lead to any permanent change in the market price. The costs experienced by the programme, however, were higher than they would have been had the credits been limited to the actual equipment costs.

For solar thermal collectors, on the other hand, a permanent increase in price of 7 per cent annually has been noticed since 2000. It is difficult to explain this change; however, an increase in raw material prices has been given as a reason. The price increase phenomenon may have been made more severe with the development of the fiscal incentive, although it is not directly proportional with tax credit amounts and it is not directly linked to the implementation of a tax credit as the main policy tool to support the renewable energy systems. It seems possible, however, that industry players are taking advantage of the tax credit rather than passing these savings to the customer. A possible solution is to communicate to the public average prices for this equipment, so that customers can make informed purchases.

Programme conclusions

The programme has been quite successful so far because of the size of the credit, increasing fossil fuel prices and the existing tax structure. This measure could be recommended to governments where citizens experience high tax pressure and are accustomed to tax credit programmes, as is the case in France.

This tax credit programme has been evaluated in comparison to similar measures in five countries (France, Belgium, Austria, Italy and Portugal) and it was determined that the French programme had the strongest effect in terms of renewable heating market growth.[24] This was in part due to the fact that it is currently the main renewable energy incentive available to citizens. In Austria, for example, the tax credits were much less successful because they compete with strong regional subsidy programmes. In Portugal, the renewable energy incentive was offered as a less-attractive alternative to a household loan interest programme, and therefore did not enjoy great success. Finally, the French programme's emphasis on qualification of equipment and installers made it more successful than the Italian programme which did not require these quality standards.

Specific features that might be considered as useful to other regions/countries are:

- Tax credits usually involve low administrative costs and hassle for both applicants and administrators.
- The technologies chosen were all mature and therefore the markets were able to respond to increased demands.
- Implementation of an accompanying information campaign aimed at both the public and all market actors, with a special emphasis on installers, is important.
- Emphasis needs to be on the development of the entire industry for each renewable energy technology.
- Emphasis on quality certification of products and installation not only leads to better and more efficient energy production, it also generates higher customer satisfaction and therefore more market growth and trust.
- A tax credit is more successful when it is the main incentive available to purchasers for a given technology. For example, in the first time period, the tax credit had to compete with other subsidies and therefore was used by only 35 per cent of buyers. However, for heat pumps during the same period, 90 per cent of buyers took advantage of the tax credit, since the credit was the main incentive available.

Analysis

The French direct tax credit appears to be a successful tax credit programme. This type of programme solves a number of problems (such as high overhead costs) but also has, as this programme has demonstrated, additional challenges (such as possible market distortions). Not surprisingly, it appears that this type of programme works best when it is accompanied by an information programme. It is interesting to note how regional programmes shifted to other efforts when the direct tax subsidy became available.

Key lessons learned

The French direct tax credit programme suggests the following key lessons:

- Different tax credit percentages for technologies should be implemented based on efficiency levels and current market price in comparison to traditional technologies.
- More information should be available about selecting efficient equipment, particularly for wood-burning appliances because so much variety exists in terms of efficiency and emissions output.

- If only the capital cost of the technology is included, market distortion may occur (for example, installers would invoice the job for 'free' but increase the capital cost in order to take full advantage of the subsidy).
- Information campaigns are important, and require two or three years for the market growth to reflect this guidance.
- In order to avoid confusion, a centralized information agency would prevent inquiring customers from receiving varying or wrong information from the tax agency, installers, sellers, energy agencies, and so on.

Norway's Household Subsidy Programme

Information about the Household Subsidy Programme in Norway was collected through a telephone interview with Even Bjørnstad of Enova SF. The focus of the interview was the initial six-week programme in 2003, for which the most information is available.

Programme description

The Household Subsidy Programme offered investment support and education measures for the implementation of heat pumps, wood pellet stoves and automated temperature controls. The subsidy covered 20 per cent of all costs including materials and installation, as long as the technology was purchased from a certified dealer and the installation was done with a certified installer. A cap subsidy value was set at 5000 NOK (approximately €550) for pellet stoves or heat pumps. Guidance was offered through a telephone hotline. The telephone hotline was staffed with engineers who were able to give good technical information and advice.

The heat pumps subsidized through the programme were air-to-air and air-to-liquid models, although most installations were of air-to-air models. After the programme, it was determined that a subsidy was no longer needed for the air-to-air models because the price has become competitive with traditional heating technologies in the Norwegian market. This was evident in the fact that heat pump sales, even without the subsidy, were greatly increased after the programme. Also, the market price of the unsubsidized heat pumps after the programme was equal to the price of the heat pumps during the subsidy programme when a subsidy was applied. Therefore, in the 2006 programme, air-to-air heat pumps were removed from the programme but air-to-liquid heat pumps are still included. As of yet, ground source heat pumps are not subsidized, but are considered a possible third 'step' in heat pump technology for the subsidy. In the 2006 programme, solar panels are included, but only a handful of applications have been made.

The subsidy was aimed at larger households with a minimum of 20,000kWh/year electricity usage; smaller consumers were discouraged from participating in the programme because they would not realize tangible savings. Despite the fact that one of the main political motivators behind the programme development was concern for low-income households, these households could only benefit indirectly as the result of lower and more stable electricity prices due to overall reduced electricity demand. This effect turned out to be negligible.

The final approved spending of the programme was approximately 80 million NOK, or about €8.8 million, which was greater than the original allocation of 50 million NOK (€5.5 million). Spending included administration fees for processing applications.

Context

Norway is a country that has abundant hydroelectric power and therefore electric heating tends to be the primary fuel used for heating in homes. This is supplemented with some traditional wood fireplaces. However, the potential for hydroelectric power expansion is limited, and when the autumn or winter is particularly cold and dry, less hydroelectric power is produced and electricity prices rise. This is what happened in the winter of 2002/2003. The Household Subsidy Programme was the politicians' immediate answer to concerns about increased heating prices, particularly for low-income households. It ran from 1 February–15 March 2003, with the intention of enabling households to choose alternatives to electricity for heating needs. The aim was to reduce electricity consumption in order to reduce electricity costs per household, as well as reduce overall demand to keep electricity prices lower for everyone. It is important to note that, although the programme was run through Enova SF, it was directly promoted and paid for by the government, unlike most Enova SF programmes which are designed and funded by the organization itself.

Results

Approximately 20,000 households received a subsidy. Of these households, 92.1 per cent installed heat pumps and 6.2 per cent installed pellet stoves.[25] Air-to-air heat pump sales were greatly stimulated by the programme, although there was no significant increase in the sales of air-to-liquid heat pumps.

Air-to-air heat pumps were already relatively similar in cost to the conventional technology available, and therefore the subsidy made them even more attractive economically and increased public awareness of the technology. In fact, heat pump sales were permanently increased, since people considered them as an attractive option even without the subsidy. For this reason, in the subsequent follow-up programme in 2006, air-to-air heat pumps have not been included because the appropriate market transformation has occurred. Pellet stoves, however, are still more expensive when compared to electric heating, and the result has been that a perpetual subsidy pattern for the technology has been

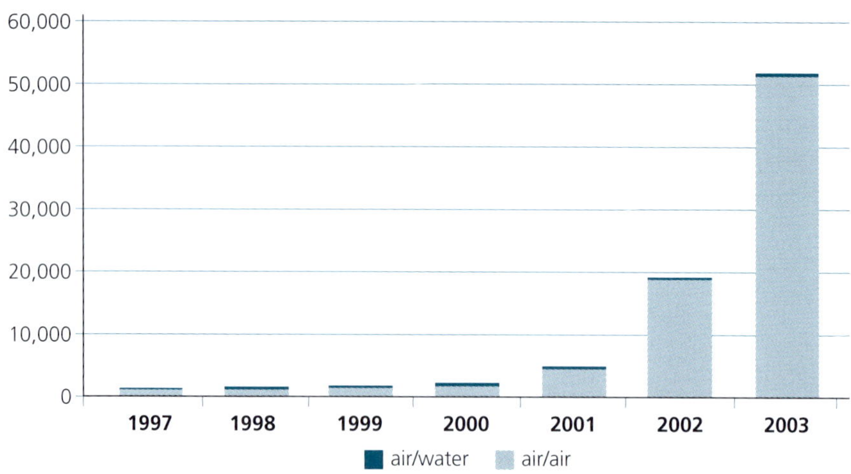

Exhibit 26 Ambient air heat pump sales before and after the Household Subsidy Programme[26]

developing. However, people were happy with their investment in pellet stoves, even if it was not necessarily a great economic investment.

Specific targets for the programme were not set – the aim was to reduce electricity consumption. However the Household Subsidy Programme was considered quite successful because:

- The programme spent more than was originally budgeted.
- Significant electricity savings were documented for the 20,000 households that participated; 5770kWh/year was saved per household on average, which was a 20.1 per cent saving of household electricity consumption, or a 32.6 per cent saving of electricity consumption for heating. This results in a total saving of 129.3GWh: 117.4GWh for heat pumps, 10.7GWh for pellet systems, and 1.2GWh for steering systems.[27]
- The majority of households were satisfied with their investment, which was determined through detailed household surveys.
- The programme helped households understand other heating options available, and a verifiable market transformation was observed for air-to-air heat pumps.
- The social cost/benefit analysis was evaluated to be marginally profitable socially, in terms of increased comfort and reduced energy costs, technology costs, grid power challenges and environmental externalities.

The main driver of the programme was the increased price of electricity during the winter of 2002/2003. The availability of the programme was an important factor in the renewable energy heating uptake, particularly for heat pumps, in response to the price increase. Growing energy awareness ranked as a distant third factor.

Design and implementation process

The dry and cold winter of 2002/2003 had significantly higher electricity rates, which prompted a debate on the cost increase for households (particularly low-income) which were largely electrically heated. Parliament decided to act in the form of this subsidy programme, although it is not clear how these three particular technologies (heat pumps, pellet stoves, steering systems) were chosen. It would have been preferable to have a formal programme of moving technologies in and out of the programme depending on the market, with the intention of avoiding the creation of permanent subsidy for any one technology, as has occurred for pellet stoves.

Enova SF is a public agency that exists to design and implement renewable energy projects. It was chosen to run the programme because it was a pre-existing structure with the personnel (40–50 people) for such programmes. Since Enova SF was relatively new (begun in 2001) at the time, the Household Subsidy Programme helped households become familiar with Enova SF. The Household Subsidy Programme required an operating budget that was approximately one-tenth of Enova SF's operating budget. The subsidy programme was always meant to be kept separate from Enova SF's programmes financed through the Energy Fund.

Enova SF was also responsible for monitoring the programme and taking corrective action. Monitoring was based on statistics regarding pre-approval and so on, and questionnaires during the programme implementation. The 2003 programme acted as a blueprint for the later programme. One corrective action was removing the air-to-air heat pumps from the programme when the subsidy was reintroduced in 2006.

The original financing was set at 50 million NOK, with no explicit decision on what proportion should be spent on administration or other expenses. The structure and value of the subsidies, including the cap values, were decided based on Swedish experience and observed market prices for the relevant technologies.

The programme was delivered mostly on the internet through ads, webpages and the online application. Media also publicized the programme. Although the programme was open to both new constructions and retrofit, the technology had to be pre-approved and bought during the programme (but could be installed later). Consequently applications were primarily for retrofit situations. Technical issues were overcome by subsidizing only equipment that met technical standards and were installed by a certified technician.

Evaluation process

One evaluation report has been completed.[28] No specific targets had been set for the 2003 programme, there was just a general aim to reduce electricity consumption. However, a social cost/benefit analysis was completed, which took into account the following social benefits:

- Saved energy costs.
- Increased comfort in households.
- Reduced environmental externalities.
- Reduced grid power challenges.
- Reduced market prices for the products.

The relevant costs were defined as:

- Administration costs.
- Investment costs.
- Pellet costs where applicable.

In this calculation, profitability is quite sensitive to electricity prices; for pellet stoves, the small difference between pellet stoves and electricity prices made it less socially profitable than heat pump technologies. The programme was also deemed successful based on household surveys.

The 2003 programme has acted as a blueprint for the 2006 programme. The evaluation of the 2006 programme is currently being developed and there is a plan to include quantified targets in the programme this time and evaluate energy behaviour.

Challenges and barriers

The primary main barrier for the technologies was, initially, cost, and this was meant to be overcome in the form of the 20 per cent subsidy for the technology and installation. In the case of air-to-air heat pumps, the market price was permanently decreased after the first year due to market growth. In fact, the winter after the subsidy, heat pump prices descended to about the same level customers were paying when the subsidy was in place, which made it possible to drop the technology from the programme. Awareness was the second barrier, which was addressed through the well-staffed hotline.

There was a time-lag issue because applications were only approved if there was funding available. There became a problem when people applied for pre-approval, and therefore the money was earmarked for use, but did not go through with the purchase. This limited the funds available for other applicants. The increased budget helped minimize this issue.

Currently, there is an issue with the Norwegian wood stove producers' agency, which complained that pellet stoves were being unfairly promoted over woodstoves. This complaint was reported to the European Surveillance Agency, arguing that pellet stoves are a base-load technology, whereas woodstoves are a peak-load technology, and therefore favour should not be given for one over the other. This complaint is now being dealt with and it highlights the need for justification of technology selection.

Programme conclusions

The programme was quite successful in Norway's specific context, particularly for heat pumps. However, it was not so successful for pellet stoves. Since markets vary greatly from country to country, it is difficult to recommend such a programme based only on Norway's experience.

It would be most transferable to countries that have a similar context to Norway in terms of electric consumption for heating. However, a subsidy programme is an easy programme to implement, and therefore it can be a good test to observe how a market reacts to a given technology.

Specific features that might be considered as useful to other regions/countries are:

- Easy web-based application procedure.
- Most suitable residential segment addressed (largest energy consumers).
- The hotline staffed by engineers.
- Approved technicians hired for installations.
- Technology supplier organizations worked in cooperation with the programme to ensure that relevant technologies were physically available in stores.
- High technical standards, which meant fewer technical problems were found with the chosen technologies. (A few brief market actors brought in imported, low-quality heat pumps for sale without the subsidy. These market actors quickly disappeared and have left some warranty issues. However, the subsidized heat pumps required an approved installer, and therefore did not have this problem.)
- A very thorough questionnaire sent to households to measure how they felt about the programme.

Analysis

This is an interesting example of a programme with a short-time frame. Most of the evidence suggests that long-term programmes are the most successful, however, this very short intervention was successful in moving the air-source heat pump market to lower prices. The programme also raised public awareness of options for responding to a tightened supply of electricity. Follow-up programmes have incorporated the lessons learned in the original programme. On the other hand, one of the original motivators for this programme was concern about the heating costs in low-income households, a concern not directly addressed by the programme.

Key lessons learned

The Household Subsidy Programme provides the following lessons:

- Entry and exit strategy must be coordinated with the market. Rumours of which technologies will be sponsored can cause market distortion. This should be minimized.
- Firm targets for installations or energy consumption reduction should be set.
- Although the evaluation for the 2003 programme assessed the behaviour of participant households, it would be useful to have a control group of households who did not participate. Also, behaviour could have been better analysed in the evaluation.

Climate Alliance of European Cities

Information about the Climate Alliance of European Cities with the Indigenous Rain Forest Peoples policy was collected through a telephone interview with Dr Dag Schulze of the Climate Alliance. Additional information was also provided by the 2007 and 2008 activity reports. Financial figures were only available in the 2007 report and therefore financial and staffing information was taken for that year. Note that the Climate Alliance also sponsors and is involved in many programmes. This summary focuses primarily on the Climate Alliance policy.

Programme description

The Climate Alliance of European Cities is a network of European cities and municipalities working on climate protection at the municipal level. These cities and municipalities have entered into a partnership with indigenous rain forest peoples of the Amazon. There are two main objectives of the alliance: to reduce greenhouse gas emissions to a sustainable level in the northern hemisphere and to support projects to conserve the rain forest and improve quality of life in the southern hemisphere through cooperative programmes. By joining, member municipalities commit themselves to targets, activity areas and measures set out by the Climate Alliance Manifesto (1990) and the Climate Alliance Declaration (2000), with the general goal of halving per capita emissions.[29] The initial target was to reduce emissions by 50 per cent (relative to 1987 emissions) in each municipality by 2010. In 2005, 2006, the Climate Alliance's general assembly enacted a new emission reduction target of 10 per cent every five years using a baseline year of 1990 in order to achieve the 50 per cent reduction by 2030 at the latest.

The Climate Alliance was officially registered in 1993 as a non-profit association with 10–20 member cities. Presently, 1400 municipalities are involved, mostly in the German-speaking areas of Europe. However, all cities in the EU are welcome to join. Members must pay €0.06/inhabitant/year to become part of the alliance, with a minimum membership fee of €180.

This programme provides municipalities with some flexibility as to how they achieve the greenhouse gas emission reductions. Some examples include:

- Climate change policy development.
- Efficiency/renewable energy projects.
- Promotional campaigns.[30]

Local authorities are encouraged to design and implement suitable local programmes as well as participate in national programmes. A best practice database is available for planners to assist in developing programmes.

There are no programmes designed by the Climate Alliance specifically for the promotion of renewable heating and cooling. However, the Climate Alliance jointly promotes various European Commission programmes involved in the residential sector, such as RES-league, which promotes solar installations on (mostly residential) city roofs.

On the other side, the indigenous partners involved in the Climate Alliance are:

- Coordinating Body for the Indigenous Peoples' Organizations of the Amazon Basin (COICA).
- International Alliance of the Indigenous-Tribal Peoples of the Tropical Forests (IAIP).

Projects and activities for the indigenous partners include renewable energy and environmental initiatives, as well as other development topics.

Context

The Climate Alliance was started by the first municipalities because of a common interest in the world's climate and the development of north-south equality. Municipalities wanted a way to get involved at a local level, and therefore the first municipalities developed the Climate Alliance and officially registered it in 1993.

Results

The overarching target for the policy when it was first designed was a 50 per cent greenhouse gas emission reduction by 2010. By 2003, some municipalities had already reduced greenhouse gas emissions by 25–30 per cent.[31] These reductions have been achieved mostly in smaller municipalities that are more tightly knit and where citizens take much pride in their community. Also, smaller municipalities generally have fewer industrial facilities, which simplifies emission reduction.

However, in 2005, it was felt that the imminent target of 50 per cent reduction by 2010 was not possible, particularly in the larger cities. As a result, the reduction deadline was extended to 2030. This deadline was selected based on estimates of emissions reductions in Munich. It was also decided that interim milestones targets were required in order to keep municipalities on track, and therefore the target of a 10 per cent reduction in emissions every five years was set.

The main drivers behind the emissions reductions in participating municipalities are most likely energy prices and reduced renewable system costs due to the subsidies offered by the programmes supported by the Climate Alliance. In fact, it was found that raising awareness about the cost of energy imports in a municipality can help trigger community interest in reducing energy consumption and emissions. General environmental awareness also assisted with the emission reductions achieved, partly due to the guidance programmes. Finally, the image and appeal of the technology was probably another factor in the success achieved.

Design and implementation process

The Climate Alliance began in 1993 as a non-profit association. Originally, it was funded only by membership fees, but now it is co-funded by the European Commission and national governments. Decisions are made by members with input from the Alliance administration and supporting federal and EU governments.

The Climate Alliance has a general assembly and an executive board made up of elected officials from local European authorities and indigenous partner representatives. The European Secretariat in Frankfurt am Main coordinates the association with support from national and regional coordination offices and contact points. In 2007, the European Secretariat employed approximately 13 people in 8.5 positions, seven of which are focused on local climate change mitigation, further supported by people working on a part-time, temporary, voluntary or internship basis. The tasks of the secretariat include supporting members by giving advice and initiating joint projects, develop guidelines and proposals for action, evaluating outcomes and coordinating campaigns for public awareness. Since 2007, a permanent secretariat has been established in Brussels in order to represent the Climate Alliance at the European Commission and the European Parliament. There are several other national and regional offices within the European region.

The expenditure for 2007 was €1,398,991, which included:

- 38.6 per cent for personnel costs and work contracts.
- 2.8 per cent for travel costs and expenses.
- 1.6 per cent building costs.
- 2.7 per cent materials.
- 2.1 per cent services.
- 7.1 per cent publications and events.
- 0.8 per cent miscellaneous costs.
- 2.2 per cent COICA.

- 16.5 per cent transmission of third-party funds.
- 25.6 per cent transfer to reserves.[32]

This was paid by earning reserves, membership dues, third-party funds and grands, COICA projects, donations and sponsoring as well as some other miscellaneous sources.

The Climate Alliance has more than 15 different websites to communicate to members and other communities. An online database is being developed of best practices to assist members in the development of effective programmes. The eClimail electronic newsletter is distributed to participating members and ministries, providing information about local climate change news, updates from the rain forest and information about upcoming events. In 2008, nine newsletters were sent out by email to 4200 participants and 15 invitations were sent out for events.[33]

More recently, the Climate Alliance has been given a contract to develop a national benchmarking system for local climate protection ('Climate Cities Benchmarking'), after the successful completion of the first phase of " 'Local Government Climate Partnership'.[34]

Evaluation process

Even with well-considered targets, performance is not easily assessed because each country has different reporting standards. It is difficult to make any conclusive statements about savings. The programme has been evaluated in some municipalities by some independent sources, however, no overall reduction measurement has been produced. The aim is to complete such an evaluation by the end of 2009.

Software has been developed to monitor savings with support from the 17 member minicipalities and districts in Germany. The software is the energy and CO_2 calculation tool ECO2-Regio from the Swiss company Ecospeed, and has been modified to suit the Alliance. The software is being used in Germany and Switzerland by 100 local and regional organizations.[35] Each municipality must pay €350 to access the software. Climate Alliance Italy is working to develop a version for the Italian municipality members of the Alliance.

An annual activity report assesses the success of some of the programmes involved and sets out plans for the next year.

Challenges and barriers

As mentioned earlier, the CO_2 emission reduction targets were initially too long-term. In 2005, member municipalities became anxious about meeting the 2010 targets in the five remaining years. Also, new municipalities were discouraged from joining because of the aggressive target and short deadline. For this reason, it was decided that the targets would be modified to a more manageable level. Using a CO_2 reduction simulation based on the city of Munich, the more realistic target of 2030 was set. Also, intermediate milestones of reducing emissions by 10 per cent every five years give municipalities more direction in accomplishing their long-term goals. This is especially important in cities because an elected city government may be accountable for a five-year term, but not usually for much longer.

Financing was initially limited to donations, some sales revenue and membership. More recently, funding has been made available through national governments and the European Commission, and therefore this barrier has been overcome.

Finally, general awareness of climate change and environmental issues has increased greatly since 1990, and therefore the Climate Alliance has grown in popularity in Europe as more and more municipalities look to make a commitment related to emissions and energy import reduction.

Programme conclusions

The Climate Alliance would be transferable to other countries or continents with similar political and economic situations as in the EU. This could be generalized to say that countries in the northern hemisphere, which tend to be more industrialized, are most suitable, since countries in the southern hemisphere tend to have other, more urgent issues to deal with than emission reduction.

Specific features that might be considered as useful to other regions/countries are:

- Supported, customizable plans can be developed for specific municipalities, with a best practice database that cities can use to develop their plan to achieve the targets.
- To date, smaller municipalities have been the most successful in achieving reduction targets. This type of programme is more effective in smaller cities, and can have a greater impact, including being a catalyst for more ambitious projects (such as 100 per cent renewable energy cities).
- Although ambitious long-term targets should be set, intermediate milestones help communities pace themselves in order to reach the long-term target. Municipal officials are more easily held accountable for shorter term targets.
- Since this is a voluntary programme, cities who do not meet their targets are not punished in any way. However, great performers can be recognized and therefore the tone of the policy is very positive.

Analysis

The Climate Alliance Policy is an interesting example of multi-country cooperation at the municipal level. The evolution of the policy statements is instructive for other programme development organizations: the need for achievable goals with intermediate milestones is clear. Willingness to adapt over time appears to have been crucial to the success of this policy.

Key lessons learned

The Climate Alliance Policy suggests the following lessons:

- A balanced approach is required to set ambitious yet reasonable targets, with intermediate milestones that help municipalities make shorter term plans for which municipal officials may be held accountable.
- Targets have no meaning if progress is not monitored.
- Local, regional, national and international authorities should work together to achieve emissions reductions.
- It is important to emphasize the costs of energy imports to a municipality in order to further justify renewable energy solutions and the associated upfront costs.

Innovative programmes

The programmes reviewed here have either been completed or have been running for several years. Although completed or established programmes are most useful for investigating best practices; new, innovative programmes suggest additional methods of promoting renewable thermal energy technologies. The following programmes were identified as new and innovative. These programmes should be reviewed for indications of success in a few years.

Funding for solar hot water systems through property taxes: Berkley, California is running a pilot programme called FIRST (Financing Initiative for Renewable and Solar Technology) to allow homeowners to finance solar PV systems through their property taxes. During the pilot phase of this project, the homeowners paid only minimal up-front costs and now will repay the loan over 20 years through an additional amount on their annual property tax. The tax obligation remains with the property if it is sold, passing a portion of the cost along to the next household to benefit from the solar system. This programme addresses the short-term planning horizon of households who typically don't remain in a home long enough to achieve a positive cash-flow. (The typical home owner stays in a home around five years in Canada.)[36] Additional benefits of this programme are attractive loan terms and rates and the fact that the loan is tied to the tax capacity of the property instead of the credit rating of the purchaser.[37] Thirty-eight households participated in the pilot programme, which is now being evaluated.[38]

Building permit fee waiver or building permit acceleration for renewable energy projects: Some communities, such as Fullerton and Huntington Beach, California, have begun programmes that waive the standard building permit fee for renewable energy projects.[39,40] These programmes appear to be only a year or two old. Various task forces and consultants have also suggested expedited building permits for projects that include renewable energy technologies, however we were unable to identify a municipality with such a programme in place.

Rental programme for solar hot water systems: In some regions, such as Ontario, Canada, homeowners are accustomed to renting, rather than owning, household hot water equipment. This presents the opportunity for third parties to rent solar hot water systems to homeowners. For example, Reliance Comfort, Canada's second largest provider of conventional rental hot water systems,[41] is currently piloting a rental solar hot water system programme targeting new home builders.[42] Utilities Kingston (Ontario) is targeting homeowners directly by offering solar hot water rental systems for $49–67/month.[43] Rental programmes address at least three barriers to solar hot water installation in Ontario: the high first cost of the technology, the history of and comfort with rented systems and the likelihood that homeowners will move before achieving a positive cash flow from a purchased solar hot water system. These pilot programmes are supported, in part, by grants from Natural Resources Canada's ecoEnergy for Renewable Heat Residential Pilot Initiative.

Buying co-ops for solar hot water systems: Buying and installing solar hot water systems in large quantities can reduce initial costs. Buying co-ops can also increase consumer confidence and reduce homeowner administrative and logistical hassles. Funded in part by grants from Natural Resources Canada, several pilot buying co-ops have recently been created in Ontario. One such group is RISE Again in Toronto.[44]

Performance contracts for solar water heating: Hawaii has a performance contract programme that allows customers to install solar water heating systems without paying upfront capital costs. Capital costs are covered by a third party and then repaid through shared savings on the homeowner's electricity bills. The legislation was written in 2005 and three utilities were approved to offer beginning mid-2007.[45] Response to the pilot testing has been strong.[46]

Renewable heating obligation: Both Germany and Ireland promote the use of renewable heating systems in the residential sector by offering direct subsidies. In Germany, the *Marktanreizprogramm* (Market Incentive Programme or MAP) has existed since 1999[47] while the Greener Homes scheme in Ireland was established in 2006.[48] Recently, both countries implemented renewable heating obligations to cover a part of the heat demand in new houses. In Germany the law on renewable heat (*EEWärmeG*) came into effect in 2009 and makes the use of renewable heat (50 per cent of the total heat demand or 0.04m^2 of solar collectors per square metre of living space) obligatory.[49] In Ireland, the 2008 national building regulations (Part L Amendment) made the use of renewables for heating mandatory in new houses (10kWh/m^2/year for thermal, 4kWh/m^2/year for electricity).[50] New homeowners in Ireland do not receive subsidies from the Greener Homes scheme,[51] however new homeowners in Germany are still eligible for MAP subsidies in Germany (subsidies are reduced by 25 per cent compared to refurbishment projects).[52]

Solar Energy Kits: Portugal has a new programme to facilitate the purchase of solar thermal 'kits'. These kits include the equipment, installation, six years of yearly maintenance, and a six-year guarantee. Because the cost of these kits will be negotiated by the government in large volumes, they should be less costly than the purchase of individual components by a household.[53]

Renewable Heat Incentive (RHI): For April 2011, the UK is developing a new programme specifically targeted at increasing renewable heat capacity. Although not all the details have been finalized, it is expected that the programme will be applicable to a wide range of renewable heating technologies at household, community and industrial levels, and will include a feed-in tariff for producing biomethane and injecting it into the gas grid. The funding will be provided through a levy on gas distributors and suppliers of other fossil fuels for heating.[54]

Australian Renewable Energy Credits for Solar and Heat Pump Hot Water systems: Some Australian purchasers of solar or heat pump hot water systems are eligible to create or assign renewable energy certificates (RECs). In general, Australian RECs are generated when electricity is generated by a renewable source; solar and heat pump hot water systems are included because of the electricity generation they offset. Certain organizations, such as those that purchase electricity for distribution to consumers, are obligated to buy a specified number of RECs. Because the value of these RECs has fallen, it is not clear how helpful this programme has been for the promotion of renewable hot water systems. This programme is currently being revised.

Notes

1   Solar cooling subsidy is available only for installations of 20m^2 or more.
2   The solar obligation can be replaced by biomass or heat pump installation.
3   Baumgartner, B., *Sonnenenergie nutzen – Kosten, Förderungen, Ersparnis*, Grazer Energieagentur, 2008.
4   Data sources: Landesenergieverein Steiermark and AEE Intec.
5   *Solar Thermal Subsidy Application Statistics,* Austria Solar, 2009.
6   *Foerdermanager* website, http://portal.foerdermanager.net, accessed June 2009.
7   IEA-RETD, *Renewables for Heating and Cooling*, Paris: OECD/IEA, 2007. Available at www.iea.org/publications/free_new_Desc.asp?PUBS_ID=1975.
8   As the standard for 'custom-built' system is not yet complete and agreed upon, no Solar Keymark is currently available for these systems.
9   Barcelona Energia Website, www.barcelonaenergia.cat, accessed May 2009.
10 IEA-RETD, *Renewables for Heating and Cooling*, Paris: OECD/IEA, 2007. Available at www.iea.org/publications/free_new_Desc.asp?PUBS_ID=1975.
11 *Barcelona Solar Ordinance*, ProSTO, Intelligent Energy Europe. Available at www.solarordinances.eu/Portals/0/STO%20template_Barcelona.pdf.
12 *Plan for Energy Improvement in Barcelona,* Ajuntament de Barcelona, 2003. Available at www.barcelonaenergia.cat/document/PMEB_resum_eng.pdf.
13 IEA-RETD, *Renewables for Heating and Cooling*, Paris: OECD/IEA, 2007. Available at www.iea.org/publications/free_new_Desc.asp?PUBS_ID=1975.

14   Droege, P. (ed.), *Urban Energy Transition – Chapter 19: Barcelona and the Power of Solar Ordinances: Political Will, Capacity Building and People's Participation*, Elsevier, 2008. Available at www.knovel.com/web/portal/browse/display?_EXT_KNOVEL_DISPLAY_bookid=2252. Also Barcelona Energy Agency website, www.barcelonaenergia.cat, accessed May 2009.

15   *Barcelona Solar Ordinance*, ProSTO, Intelligent Energy Europe. Available at www.solarordinances.eu/Portals/0/STO%20template_Barcelona.pdf.

16   IEA-RETD, *Renewables for Heating and Cooling*, Paris: OECD/IEA, 2007. Available at www.iea.org/publications/free_new_Desc.asp?PUBS_ID=1975.

17   REFUND+ website, www.energies-renouvelables.org/refund, accessed May 2009.

18   The cap is for the maximum amount of investment that may benefit from the tax measure. The cap value applies to the whole taxable period, meaning that any remaining investment room may be passed on from year to year until the taxable period ends.

19,20 Observ'ER, *REFUND+ Quantitative and Economic Assessment of the Direct Fiscal Measures: Study of the French Case*, Angers, France: ADEME, 2007. Available at http://ftpnrj.free.fr/refund+/French_economic_report.pdf.

21,22,23 Source: Observ'ER

24   Observ'ER, *REFUND+ Comparative Analysis of National Feedbacks and Recommendations*, Angers, France: ADEME, 2008. Available at http://ftpnrj.free.fr/refund+/WP4%20cross%20country%20analysis.pdf.

25   Dahlborn, B. et al,*Changing Energy Behaviour: Guidelines for Behavioural Change Programmes*, IDEA, 2009.

26,27,28 Bjornstad, E. et al, *Evaluering av tilskuddingsordningen til varmepumper, pelletskaminer og styringssystemer. Nord-Trondelagsforskning*, 2005.

29   Climate for Change website, www.climateforchange.net/index.php?id=23&L=0, accessed June 2009.

30   Climate Alliance website, www.klimabuendnis.org/home.html?&L=1download%252Fsatzung-2007-en.pdf, accessed March 2009.

31   *Statement of the Climate Alliance of European Cities with Indigenous Rainforest Peoples.* Climate Alliance, December 2003. Available at http://unfccc.int/resource/docs/2003/cop9/stmt/ngo/004.pdf.

32   *2007 Activity Report and 2008 Planning*, Frankfort am Main: Climate Alliance, 2008. Available at www.klimabuendnis.org/uploads/media/activity-report-planning_2008_en.pdf.

33,34,35 *2008 Activity Report and 2009 Planning*, Frankfort am Main: Climate Alliance, 2008. Available at www.klimabuendnis.org/uploads/media/Climate_Alliance_report2008-planning2009short.pdf.

36   Canada Mortgage and Housing Corporation, 'Mobility characteristics of Canadian households', *Research Highlights, Socio-Economic Series*, no 1. Available at http://dsp-psd.pwgsc.gc.ca/Collection/NH18-23-1E.pdf.

37   Berkeley Lab and the Clean Energy States Alliance, *Case Studies of State Support For Renewable Energy, Property Tax Assessments as a Finance Vehicle for Residential PV Installations: Opportunities and Potential Limitations*, 2008. Available at www.cleanenergystates.org/library/Reports/LBL_property-tax-finance_Feb08.pdf.

38 City of Berkley, *Berkley First*, www.ci.berkeley.ca.us/ContentDisplay.
    aspx?id=26580, accessed July 2009.

39 City of Fullerton, *Administrative Services,* www.ci.fullerton.ca.us/depts/admin_serv/
    permits.asp, accessed July 2009.

40 The City of Huntington Beach, *Residential Program,* www.ci.huntington-beach.
    ca.us/residents/green_city/residential_programs, accessed July 2009.

41 Government of Canada, *Renewable Heat Program's Residential Pilot Initiative
    Project Summaries.* Available at www.ecoaction.gc.ca/ECOENERGY-ECOENERGIE/
    heat-chauffage/projectsummaries-resumesprojets-eng.cfm, accessed July 2009.

42 Reliance Home Comfort, *Solar Water Heating from Reliance,* www.
    reliancehomecomfort.com/solar/pdf/SolarWaterHeater_builder.pdf, accessed July
    2009.

43 Utilities Kingston, *Solar Domestic Hot Water System Rentals*, www.utilitieskingston.
    com/heaters/solar.html, accessed July 2009.

44 Our Power, *Toronto – RISE Again*, http://riseagain.ourpower.ca/Portal.
    aspx?portalid=1, accessed July 2009.

45 Cillo, P., *Pay As You Save® (PAYS®).* Policy Leadership Institute, 2008, www.
    carseyinstitute.unh.edu/
    documents/P.%20Cillo%20EEI%2008%20PLI%20III%20Carsey%20Presentation.
    pdf, accessed July 2009.

46 Hawaiian Electric Company, Solar Energy, http://heco.com/portal/site/heco/menuitem.
    508576f78baa14340b4c0610c510b1ca/?vgnextoid=57e9202fe559b110VgnVCM10
    00005c011bacRCRD&vgnextfmt=default&cpsextcurrchannel=1, accessed July 2009.

47 Langniß, O. et al, *Evaluierung von Einzelmaßnahmen zur Nutzung erneuerbarer
    Energien (Marktanreizprogramm) im Zeitraum Januar 2004 bis Dezember 2005*,
    Bundesministeriums für Umwelt, Naturschutz und Reaktorsicherheit, 2006.

48 International Energy Agency, *Energy policies of IEA countries – Ireland 2007 review*,
    2007.

49 *Das Erneuerbare-Energien-Wärmegesetz im Überblick*, Bundesministeriums für
    Umwelt, Naturschutz und Reaktorsicherheit, 2008.

50 Government of Ireland, *Technical Guidance Document L – Conservation of Fuel
    and Energy – Dwellings*, Dublin: Stationery Office, 2008. Available at www.
    environ.ie/en/Publications/DevelopmentandHousing/BuildingStandards/
    FileDownLoad.16557.en.pdf.

51 Sustainable Energy Ireland, *Greener Homes Scheme Phase III (Existing Dwellings),
    Application Guide*, version 3.2. Available at www.sei.ie/Grants/GreenerHomes/
    Homeowners/How_to_Apply/Greener_Homes_Application_Guide.pdf.

52 *Richtlinien zur Förderung von Maßnahmen zur Nutzung erneuerbarer Energien im
    Wärmemarkt – Vom 20, 2009*, Bundesministeriums für Umwelt, Naturschutz und
    Reaktorsicherheit, 2009.

53 IEA, *Global Renewable Energy Policies and Measures Database,* www.iea.org/
    textbase/pm/?mode=re&id=4324&action=detail, accessed August 2009.

54 Department for Business Innovation and Skills (UK), *Renewable Heat Incentive (RHI)*,
    2009, www.berr.gov.uk/energy/sources/renewables/policy/renewableheatincentive/
    page50364.html, accessed August 2009.

8
Part 1 Glossary

Abbreviations

ADEME: Environment and Energy Management Agency (France)
BAFA: Federal Office for Economy and Export Control (Germany)
CEN: European Committee for Standardization
COICA: Coordinating Body for the Indigenous Peoples' Organizations of the Amazon Basin
DSM: demand-side management
EU-CERT.HP: European Certified Heat Pump Installer
FIRST: Financing Initiative for Renewable and Solar Technology (California)
GDP (PPP): gross domestic product adjusted for purchasing power parity
GDP: gross domestic product
IAIP: International Alliance of the indigenous Tribal Peoples of the Tropical Forests
IEA: International Energy Agency
LUEC: Levelized Unit Energy Cost
MAP: Market Incentive Programme (Germany)
OECD: Organisation for Economic Co-operation and Development
REDI: Renewable Energy Development Initiative
REHC: renewable energy heating and cooling
SEI: Sustainable Energy Ireland

Terms

Achievable potential: An estimate of displaced consumption of energy from traditional sources that can be realistically achieved given institutional, economic and market barriers. Achievable potential is typically some fraction of economic potential.
Air-source heat pump: A heat pump which uses the outside air as a heat sink or source.
Air-to-air heat pump: A heat pump which uses air as both a heat source and heat sink.
Air-to-liquid heat pump: A heat pump system in which a liquid is circulated in the conditioned space to provide heating or cooling. Air is used as the heat source or sink.

Biomass: Vegetable matter used as a source of energy. Biomass materials examined in this study include pellets, wood and wood waste.

Carrots: See Economic incentives.

Economic incentives: Grants, loans, rebates and/or tax credits used to encourage the adoption of renewable technologies. Also referred to as 'carrots'.

Economic potential: An estimate of displaced consumption of energy from traditional sources if all cost effective REHC technologies were implemented. Economic potential is typically some fraction of technical potential.

Free ridership: A measure of the number of people who would have installed a given technology in the absence of a programme.

Full market stage: At the full market stage, the public is knowledgeable about the target technology and its benefits, and the technology can be easily obtained and installed.

Geothermal: Power generated from heat originating in the interior of the Earth including natural steam, warm water, warm rocks or earth.

Gross domestic product adjusted by purchasing power parity: An adjustment to GDP which takes into consideration the price of goods traded within a country (based on a specific 'basket of goods') rather than considering only the export value and exchange rate of the country's currency.

Gross domestic product: The total market value of all goods and services produced in a country in a given year.

Ground-source heat pumps: Heat pumps which uses the Earth as both a heat source and heat sink.

Guidance: See Information and marketing.

Heat pump: A device which uses mechanical energy to transfer heat from one place (the 'source') to another (the 'sink').

Information and marketing: A type of renewable energy promotion programme that includes public education and awareness campaigns, quality assurance standards, support resources and training. Also referred to as 'guidance'.

Initial deployment market stage: At the initial deployment market stage, clients may be largely unaware of the promoted technology, technology may not be readily available or installation and maintenance support may be difficult to obtain. Programme activities may include demonstration and pilot projects and training for retailers and installers.

Levelized unit energy cost: The life cycle cost of producing one unit of useful energy by means of a given technology. Used to compare the attractiveness of various energy supply options.

Mass market stage: At the mass market stage, target technologies are becoming known to the public and are available through retailers or installers. Programmes may include 'carrot' and 'guidance' activities.

Multi-unit dwellings: A residential building intended to house more than one family, such as an apartment building or row house.

Passive solar thermal systems: Solar thermal systems that do not involve moving components such as pumps. These are distinct from 'active solar thermal systems' that typically involve moving components.

Pellets: Pulverized biomass materials compressed into pellet form in order to ease transport and handling and combusted to provide heat.

Portfolio planning: Programme planning phase encompassing initial problem definition, setting objectives, identifying intervention points, identifying stakeholders and factors and selecting instruments.

Programme design: Programme phase preceding implementation and ideally based on the portfolio planning phase. May be undertaken by the funding agency or a third party.

Programme evaluation: An assessment of an ongoing or completed programme against objectives. May include an assessment of problems encountered, relevance, implementation issues and cost-effectiveness.

Programme implementation: Phase in which the programme is delivered to end users. This phase may be undertaken by the funding agency or a third party.

Programmes: In this report, this term is used to refer to any organized effort to encourage the use of renewable heating and cooling technologies. It includes efforts that may be referred to elsewhere as 'policies'.

Regulations: Building codes, and/or minimum standards requiring the use of renewable technologies. Also referred to as 'sticks'.

Spillover: The improvements that occur as the result of programme influence but are not due to programme participation.

Sticks: See Regulations.

Technical potential: An estimate of displaced consumption of energy from traditional sources if all technically viable REHC technologies were implemented immediately in every appropriate application, regardless of cost.

PART 2

Best Practices Guide

9
This Guide

Preface

This guide provides information and advice on proven best practices to support the deployment of renewable energy technologies for heating and cooling (REHC) in the residential sector. The advice is aimed at policymakers and others in governments, agencies and utilities involved in four programme phases: portfolio (strategy) development; programme design; programme implementation; and programme evaluation.

The information in this guide is drawn from real world experience with the programmes documented in Part 1. The research covers the ten IEA-RETD[1] member countries (Canada, Denmark, France, Germany, Ireland, Italy, Japan, the Netherlands, Norway and the UK), as well as Austria, Spain and the US. From the portfolios of programmes implemented in these countries, 12 successful REHC programmes were selected. Interviews were conducted with programme personnel and publicly-available information was collected to produce detailed descriptions of these programmes. These descriptions were then analysed to develop the best practices found here. Because the information is drawn from real examples (bottom-up approach), the guide does not present a systematic menu of options for overcoming all potential barriers. Rather, it is hoped that these lessons learned from real programmes can complement other sources of guidance.

This guide contains four flowcharts, one for each programme phase. To use this document, turn to the flowchart that matches the programme phase of interest. Each flowchart is organized by market stage and programme barrier. Identify the market stage and programme barrier of interest, then turn to the suggested tabs to find the best practices most likely to be applicable to your situation. The overall framework for the presentation of the best practices and explanations of the terminology can be found in Section 9.1.

9.1 Introduction

This guide provides best practice information and advice on portfolio development and design, implementation and evaluation of programmes supporting residential renewable energy for heating and cooling (REHC) technologies.

Policymakers, programme developers and implementing and evaluating bodies at municipal, regional, national and multinational levels form the target audience for this advice. As explained further in Section 9.4, the information in this guide is based on Part 1's *Best Practices in the Deployment of Renewable Energy for Heating and Cooling in the Residential Sector.*

9.2 How to use this guide

A flowchart has been developed for each programme phase covered in this document:

- Portfolio planning (Exhibits 2 and 3).
- Programme design (Exhibits 4, 5 and 6).
- Programme implementation (Exhibits 7 and 8).
- Programme evaluation (Exhibit 9).

To use this guide, turn to the flowchart focusing on the programme phase you are developing. Identify the appropriate market stages and barriers for your target region. The flowchart will identify the best practices that are likely to be applicable to your situation. Then turn to the appropriate Best Practice (10.1–10.32). If you are looking for more general information, these best practices can also be used as a collection of 'lessons learned' to help you learn more about general success factors and pitfalls to be avoided.

Programme phases, market stages, technologies and barriers are discussed in Section 9.3. If you are unfamiliar with these terms, you should read this section before referring to Exhibits 2–9. If you need help understanding other terms and abbreviations used in this guide, refer to the Part 2 Glossary in Chapter 11.

Please note that, although you will find 32 best practices in this guide, this is not exhaustive. These best practices are based on the 12 programmes examined in depth in Part 1.

9.3 Scope of the guide

The scope of this guide is defined in several dimensions: programme design phase, market stage, barrier faced, applicable technology and applicable sector. Exhibit 1 pictorially represents the dimensions of this project. The dimensions inside the dashed box – programme phase, market maturity stage and barriers to deployment – were used to organize the best practices in this guide. Each dimension is discussed below.

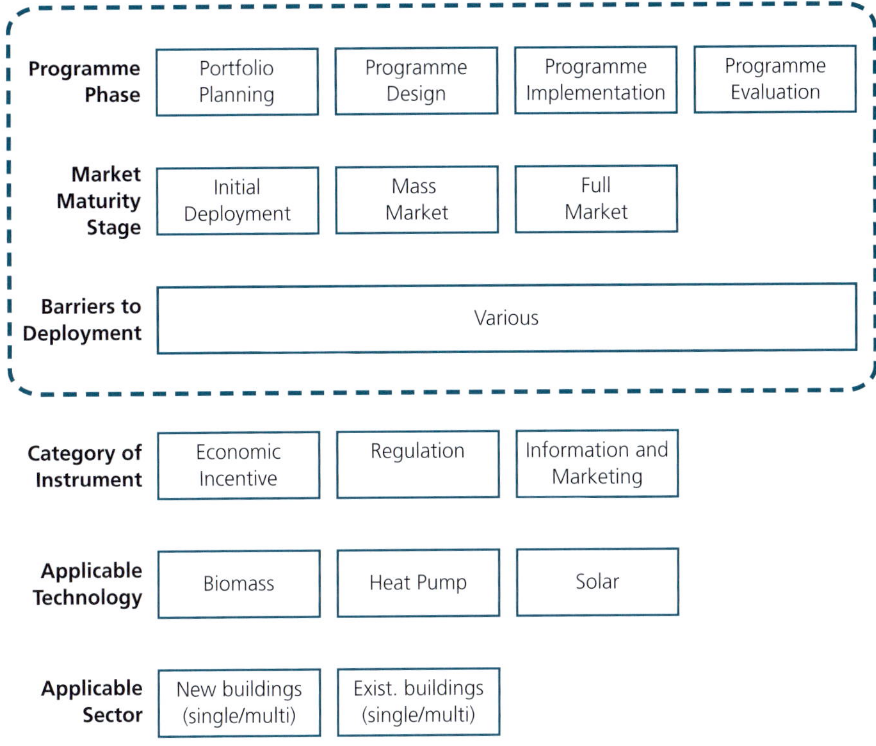

Exhibit 1 Project scope

9.3.1 Programme phases

REHC programme development can be divided into four phases:

Portfolio planning encompasses initial problem definition and goal clarification, setting objectives, consideration of options and alternatives, identifying intervention points, indentifying stakeholders and conducting consultations (if necessary) and selecting instruments.

Programme design can be done by the funding agency or a third party. Ideally, the programme design is based on the work done during the portfolio planning phase.

Programme implementation can also be done by the funding agency or a third party. During this phase, the programme is implemented.

Programme evaluation closes the loop by assessing progress against objectives, and reassessing both programme logic and process. The programme is then modified as required. Evaluations are best performed by independent bodies.

9.3.2 Market maturity stages

Programmes designed to develop REHC markets can be categorized in the following manner:

Initial deployment programmes are appropriate for a market that is in initial stages. In this type of market, the public is largely unaware of the technology, the technology is not readily available and installation and maintenance support is difficult to find. REHC programmes for this market stage include demonstration and pilot projects. Training for retailers and installers is also appropriate.

Mass market programmes are aimed at a market that has passed through the initial deployment stage. The technology is becoming better known to the public and is being sold by retailers; and competent installers can be readily found. Appropriate programmes for this market include 'carrot' programmes such as economic incentives and 'guidance' programmes to educate both the public and industry members. In some specific situations, such as under certain political conditions, 'stick' measures may be appropriate, although these are more typically found under 'full market' conditions.

Full market programmes target well-developed markets. In these markets, the public is knowledgeable about the technology and its benefits, the technology can be easily obtained and an adequate number of well-trained installers are in place. 'Carrot' measures may still be appropriate, but 'stick' measures such as mandatory installation regulations may be appropriate for bringing non-participants into the market.

Note that it is not unusual for different REHC technologies to be in different market stages in a given geographical area.

9.3.3 Barriers

REHC technologies face a variety of barriers to wide-spread use. These barriers can be categorized in a number of ways. One method uses five 'A's:

Acceptability: Some REHC technologies may not (yet) be as easy to use as conventional technologies. Maintenance costs might be higher, reliability might not be as good, or use may not be as convenient.

Accessibility: Laws, regulations or codes may make REHC installation or use difficult. For example, building codes may require expensive inspections of solar water collectors on roofs.

Affordability: Many REHC technologies are more expensive to install than conventional technologies. Financing for REHC technologies may be harder to acquire than for conventional technologies.

Availability: REHC technologies may have limited availability or lack qualified installers

Awareness: The benefits of REHC technologies may not be well-known, or potential users may not know how to obtain the technology.

Programme best practices vary not only by programme development phase and market stage, as described above, but also by barrier

9.3.4 Instrument categories

Methods of involving public and industry in REHC programmes can be divided into three categories: economic incentives; regulations; and information and marketing. Another common method of categorizing support instruments is with the descriptive terms 'carrots', 'sticks' and 'guidance'. These categorization methods often, but not always, correspond to each other in a one to one manner.

Economic incentives (often called 'carrots') include grants, loans, rebates and tax credits. Economic disincentives such as fines, on the other hand, would be 'sticks' (see below).

Regulations (often called 'sticks') include building codes and laws requiring minimum installations. In some cases, regulations can be 'carrots' in that they make installing a REHC technology easier.

Information and marketing (often called 'guidance') includes public awareness and education campaigns, quality assurance standards, support resources and training.

9.3.5 Target technologies

The target technologies for this guide are:

* Active solar thermal (for water or space heating).
* Biomass (pellets, wood and wood waste).
* Ground-source heat pumps.
* Air-source heat pumps (air-to-air, for homes with ducts; and air-to-liquid, for homes with hydronic heating).

9.3.6 Target sectors

In this guide, 'residential sector' includes single-family houses as well as the multi-unit housing sector. Best practices addressing both new and existing homes are included.

The difference between the construction of new houses and the refurbishment of existing buildings is important. It is easier to introduce, for instance, minimum requirements on REHC usage for the construction of new houses, while addressing the existing sector often requires offering economic incentives. Because there are more existing homes, the potential for REHC deployment in the existing housing stock is much larger than that in the small number of new buildings.

9.4 Source of information provided in this guide

The information provided in this guide was developed through an IEA-RETD project which analysed programmes supporting REHC deployment in the residential sector. Full details on this project can be found in Part 1.

This project built on an earlier joint IEA/IEA-RETD project which produced, in 2007, a report entitled *Renewables for Heating and Cooling – Untapped Potential*. This earlier report provided an overview of renewable energy for heating and cooling technologies and discussed the applications, market status, and research needs and priorities for these technologies. The report also included a study of programmes supporting REHC deployment in 12 Organisation for Economic Co-operation and Development (OECD) countries and a discussion of good practices and lessons learned for each type of policy instrument.

The first step in the project was to review framework conditions and policies in ten RETD member countries: Canada, Denmark, France, Germany, Ireland, Italy, Japan, the Netherlands, Norway and the UK. In addition, Austria and Spain were also profiled due to their success with renewable energy; the US was included due to both programme success and variety; and an abbreviated summary of the rapidly developing solar thermal energy market in China was also prepared. The findings were collected in country summaries, providing background information for further investigation and evaluation of promising programmes (Chapters 6 and 7). They also allowed a comparison between countries to identify leaders in REHC implementation.

Topics investigated in the country analysis were the general national energy context, REHC technology market status, and significant residential REHC programmes. Furthermore, data on resource availability, GDP (PPP) per capita, costs of conventional energy sources, installed REHC capacities or annual installations and relative costs of REHC technologies were also gathered. Where possible, the relative progress each country has made in using REHC technologies was compared.

The following criteria were used to choose programmes identified during the country analysis for further analysis:

- Evidence that programme likely caused a significant increase in renewable technology uptake based on either:
  - programme evaluation reports;
  - country or region has either high installed capacity or a recent high rate of installed capacity under circumstances that are not all favourable. Under these conditions, it is likely that programmes are responsible for the high uptake.
- Programme has been unusually stable and continuous over a period of time.
- Programme has unique features not found in other programmes/policies being studied.

- Enough historical data is available to allow a determination of how well the programme is working and what lessons have been learned.
- Programme is of sufficient scale (time, region covered or budget) to have made an impact.
- Programme selected from the most successful countries, as determined in during the country research phase of the study.
- Programme provides diversity among programmes/policies chosen, in terms of technology(ies) targeted, type of programme and dwelling types and markets targeted.

The long list of programmes identified during the first phase of the research was screened using the above criteria, producing the following final programme list:

- The Energy Action Plan of Upper Austria.
- The Climate Alliance of European Cities.
- The German Market Incentive Programme (MAP).
- Regional Austrian financial support instruments in Salzburg, Steiermark and Vorarlberg (total of three programmes).
- Regional Austrian solar information campaigns in Steiermark and Tirol (total of two programmes).
- The Household Subsidy Programme in Norway.
- The Tax Credit Programme in France.
- The Barcelona Solar Thermal Ordinance in Spain.
- The European Solar Keymark.

Once these programmes were chosen, in-depth information was collected from publicly-available data and through interviewing persons responsible for the design and implementation of the selected policies and programmes.

The detailed programme profiles were studied to identify best practices. A practice or lesson learned was considered a best practice if:

- It was found to be widely used in successful programmes.
- It was found to be used, often more than once, in countries that have been particularly successful at promoting renewable technologies.
- It was found in a programme that has been particularly successful and appears to have been essential to that programme's success

In total, 32 best practices were identified and described.

Note

1   See www.iea-retd.org/ for additional information about the RETD.

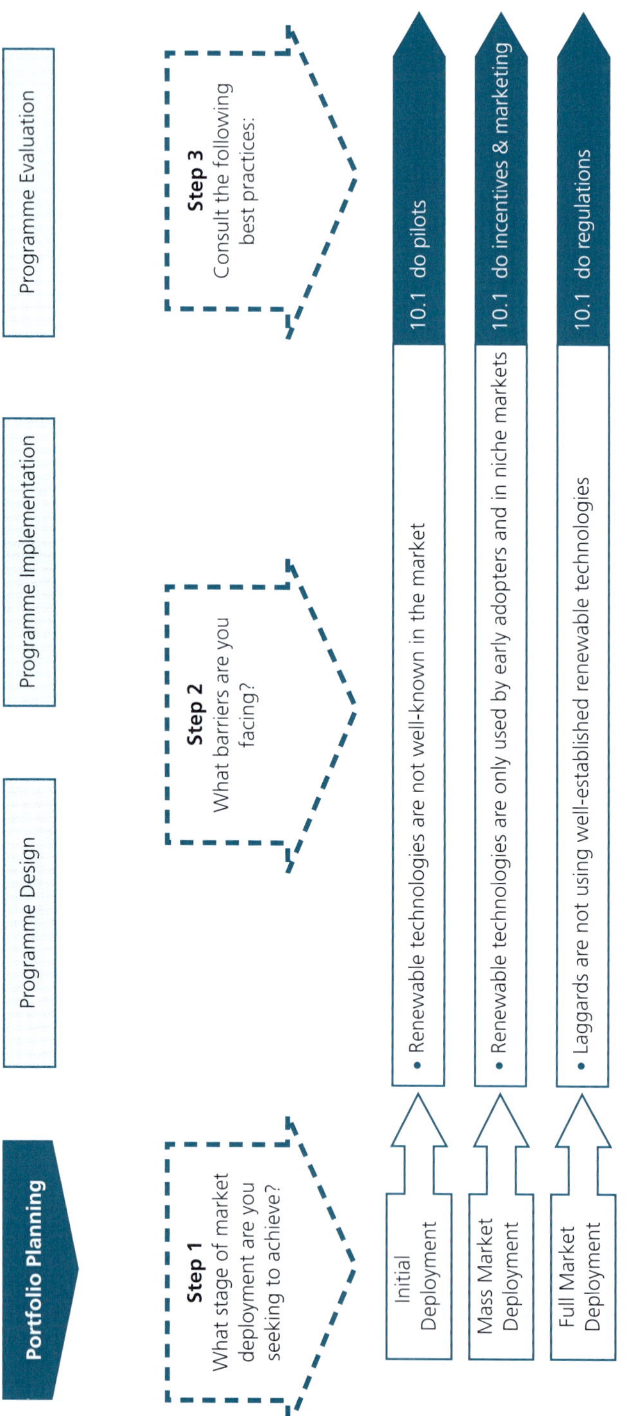

Exhibit 2 Portfolio planning flowchart part 1

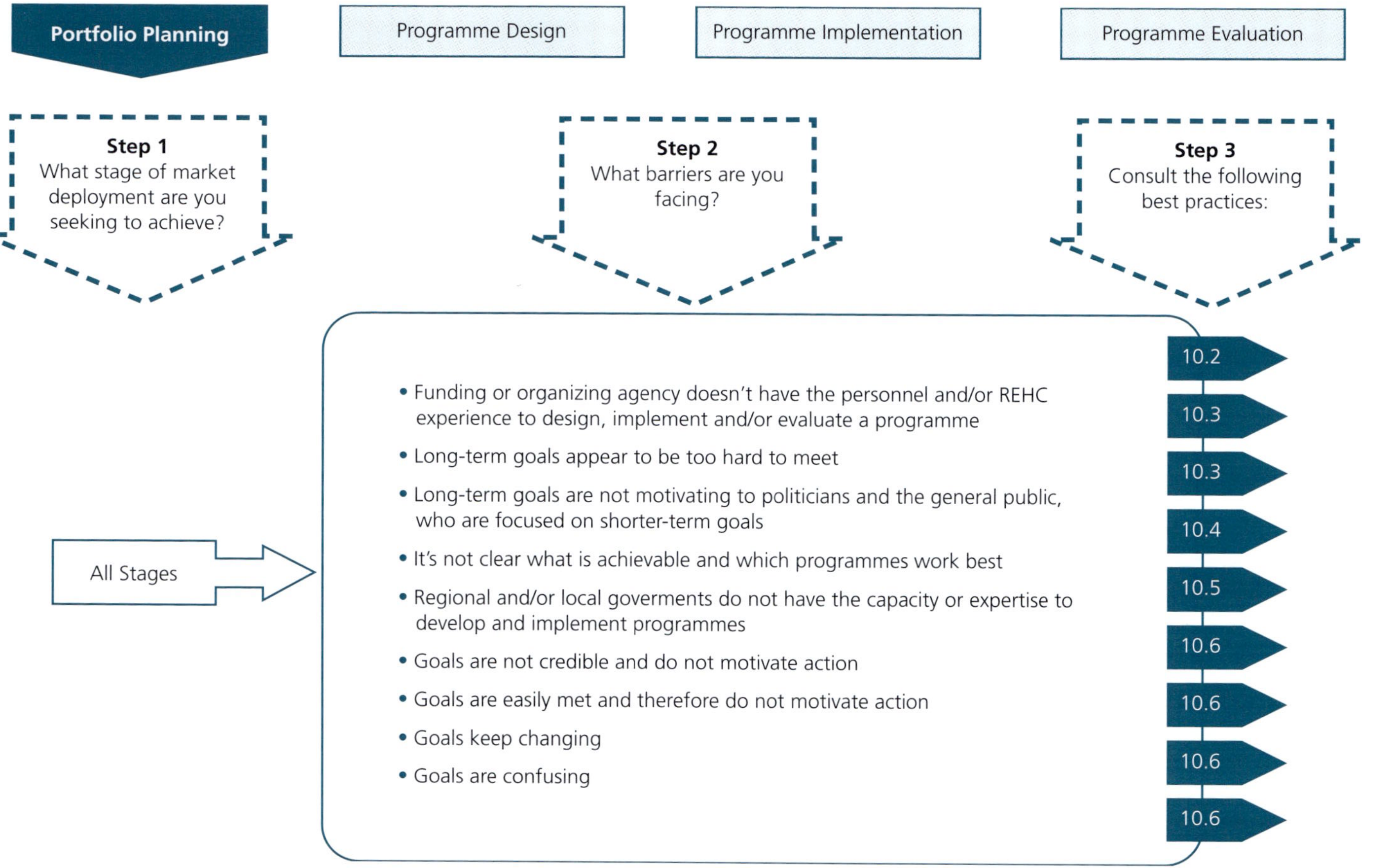

Exhibit 3 Portfolio planning flowchart part 2

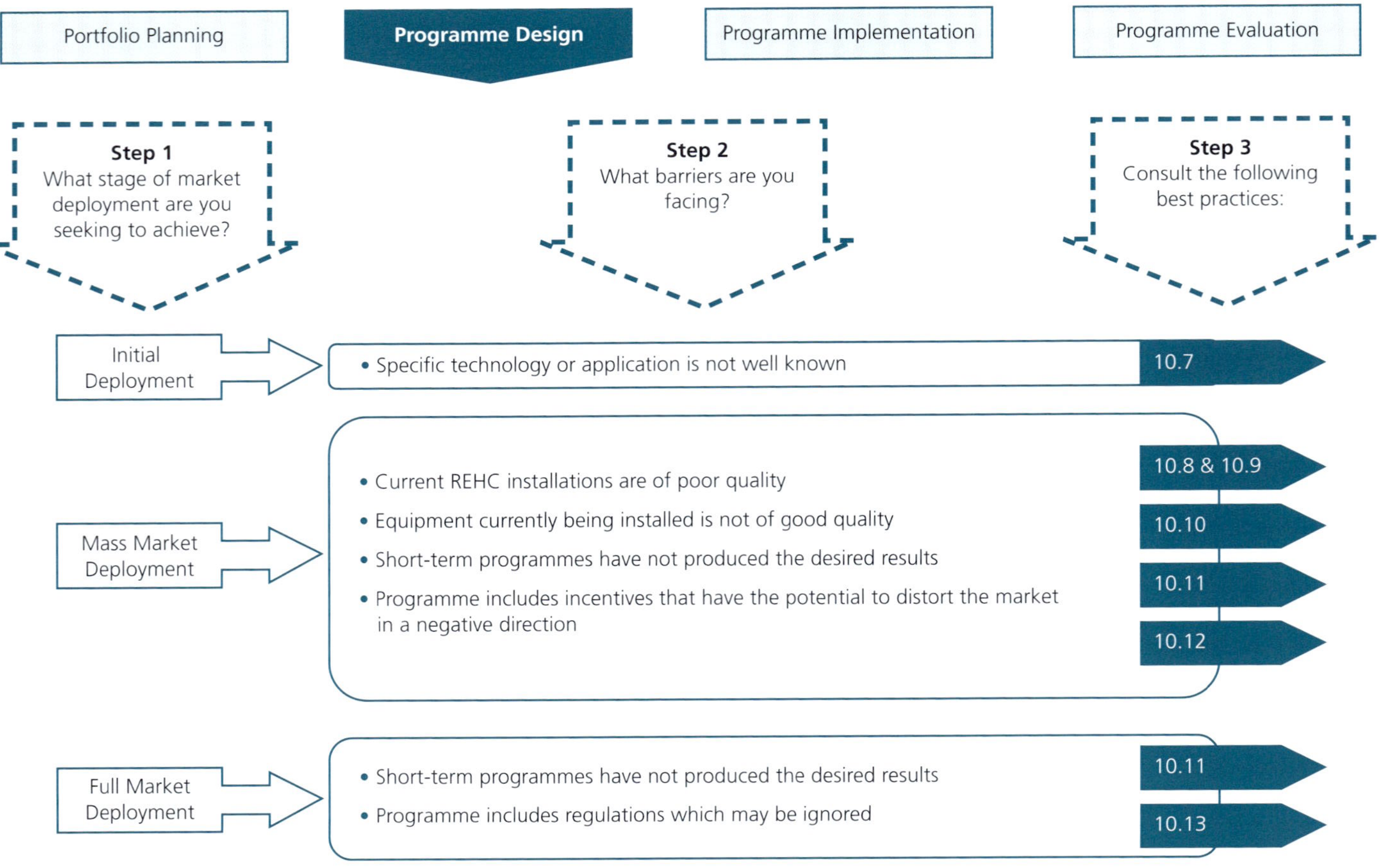

Exhibit 4 Programme design flowchart part 1

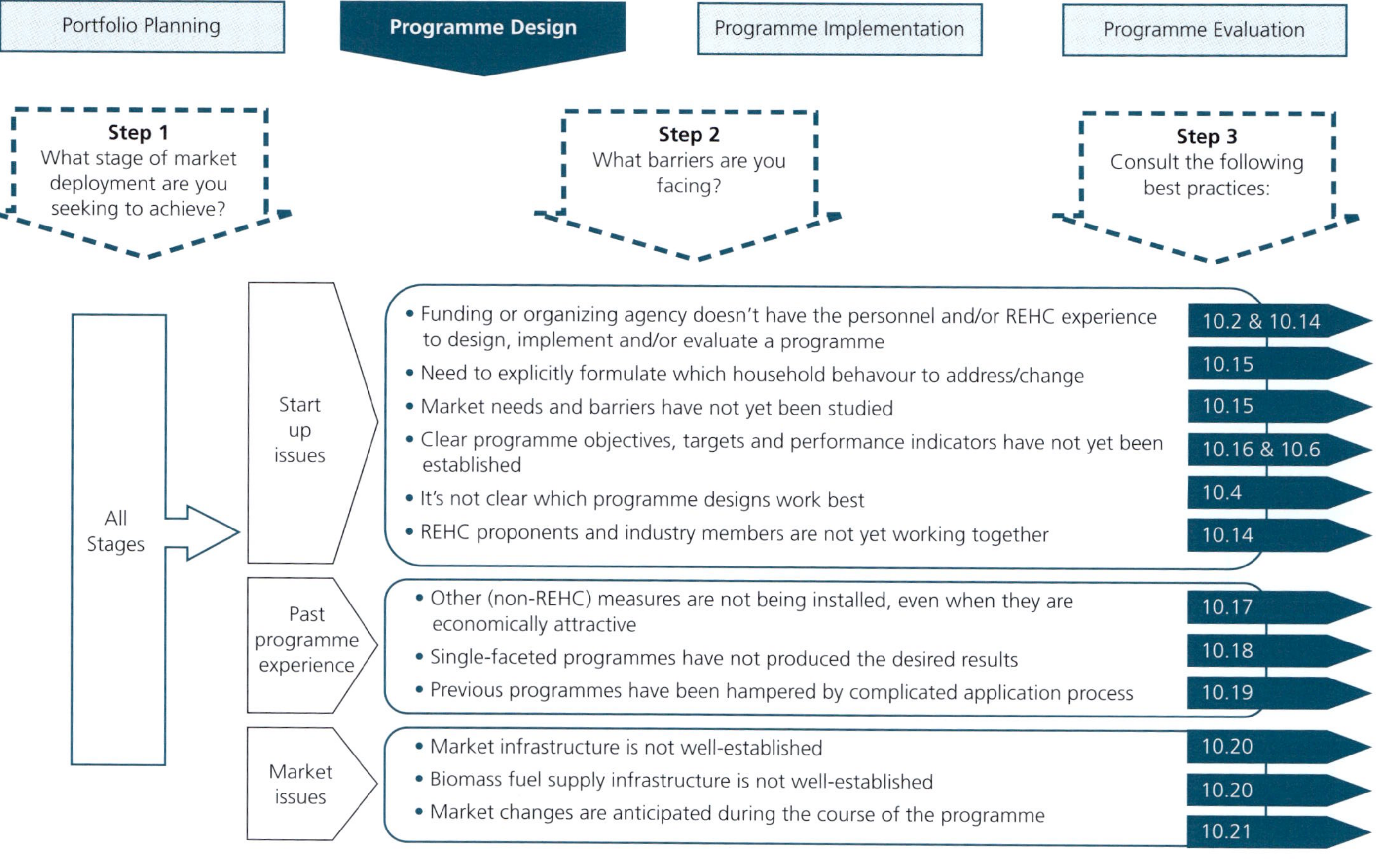

Exhibit 5 Programme design flowchart part 2

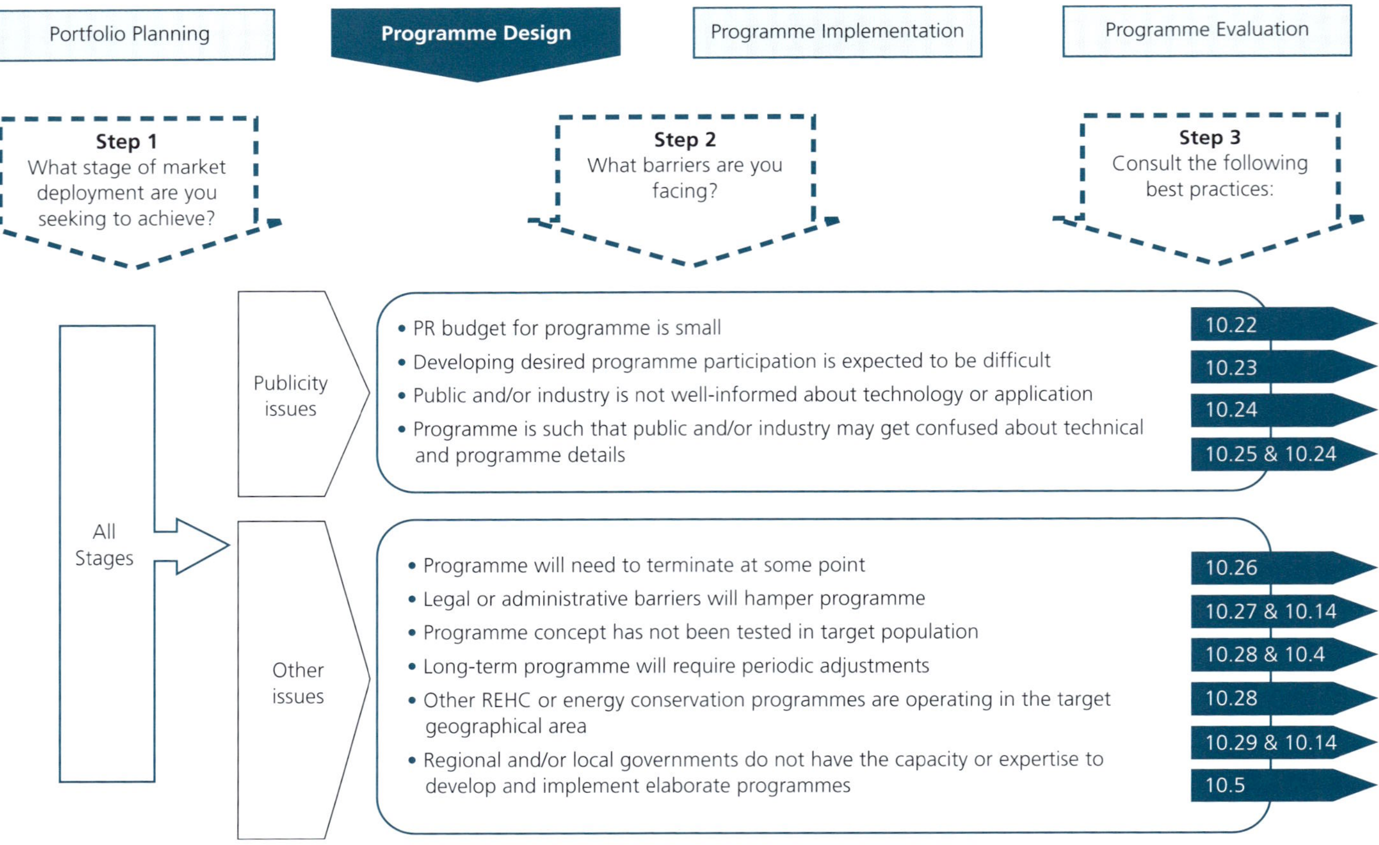

Exhibit 6 Programme design flowchart part 3

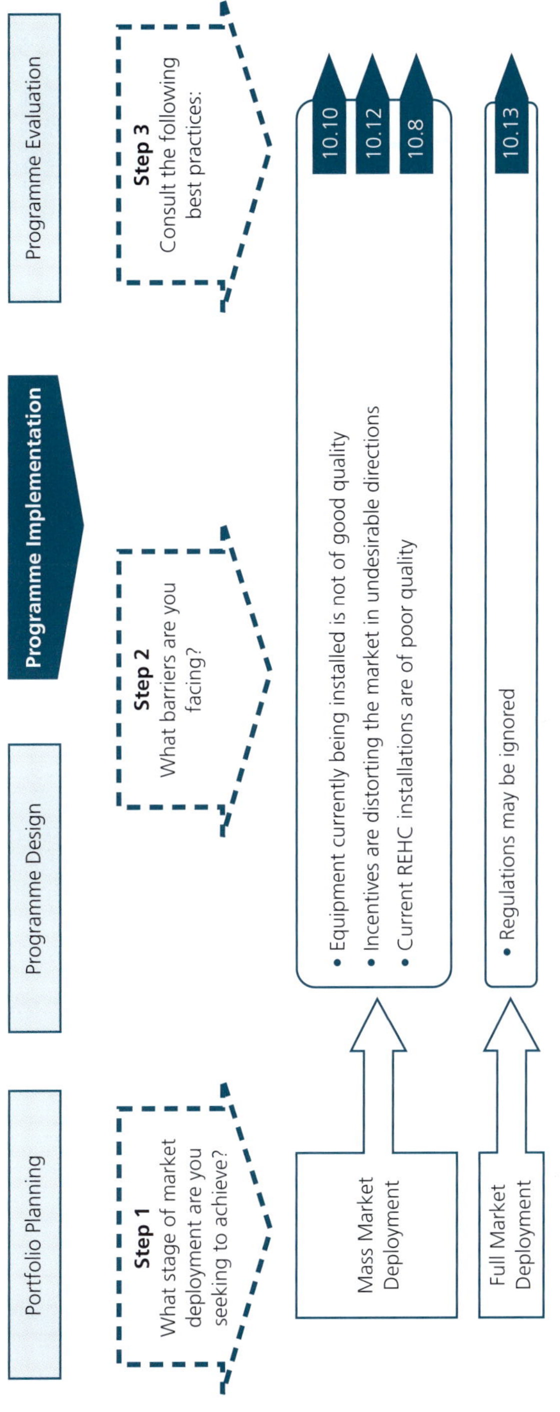

Exhibit 7 Programme implementation flowchart part 1

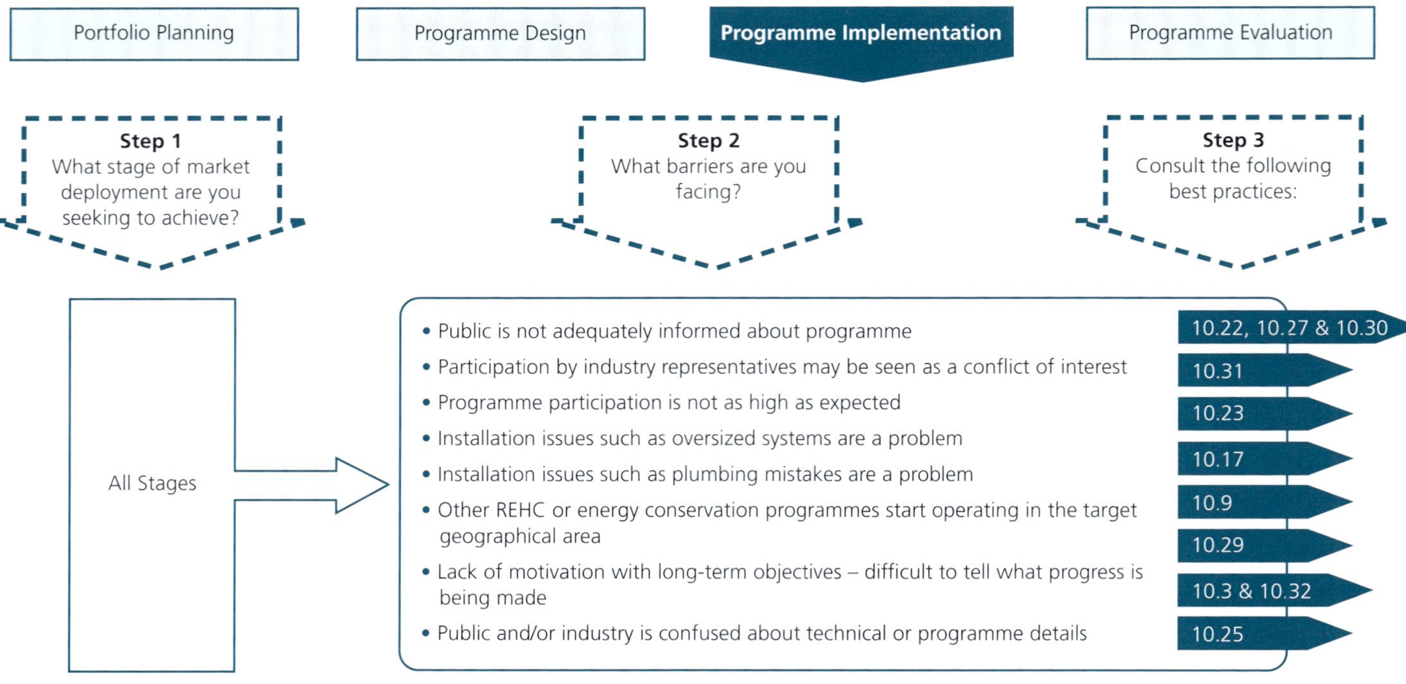

Exhibit 8 Programme implementation flowchart part 2

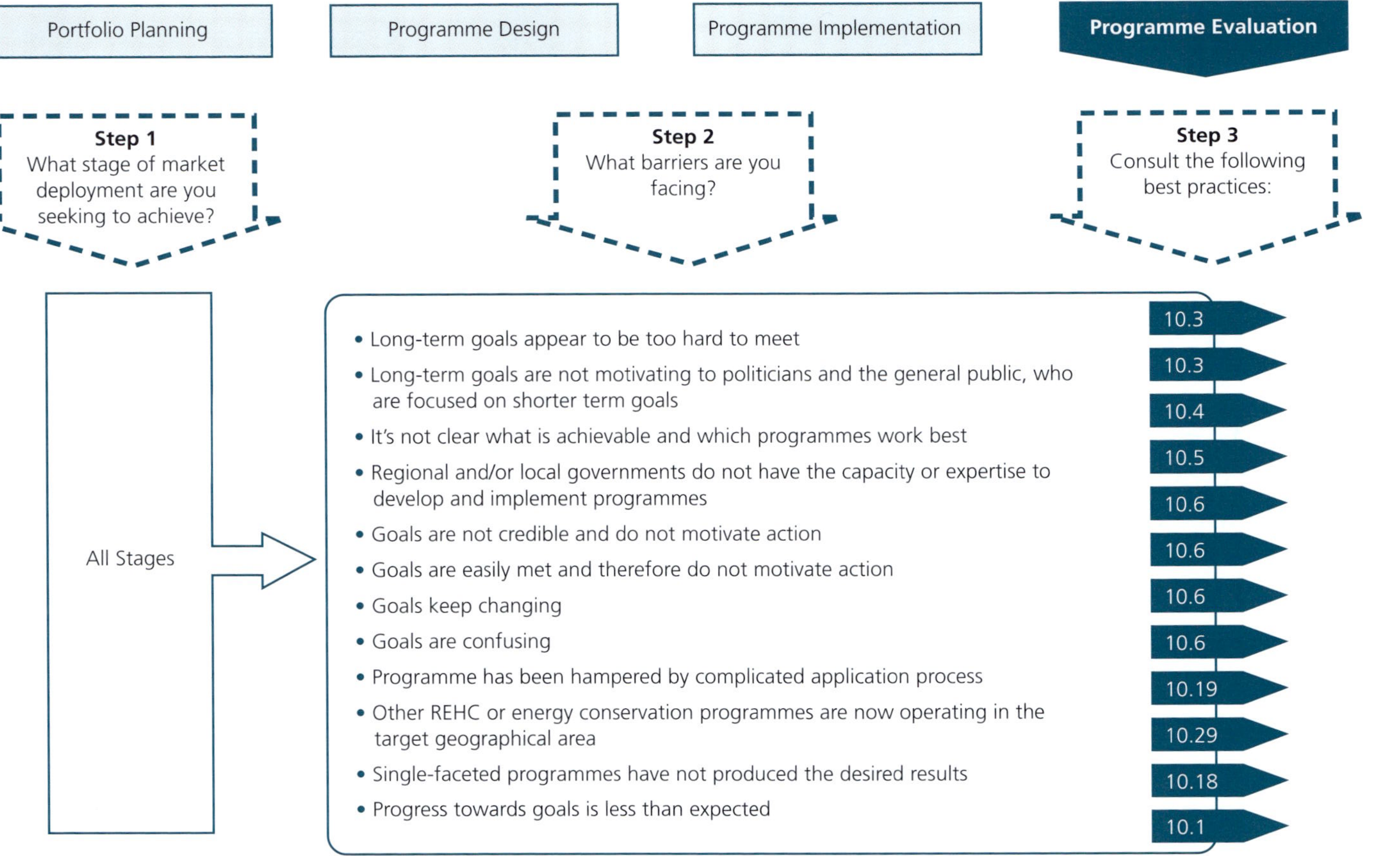

Exhibit 9 Programme evaluation flowchart

10
Best Practices

10.1 Best Practice 1: Establish plans and programmes to achieve the established targets. These programmes should address all market participants and should be long-term

Programme phase(s)

- Portfolio planning.
- Programme evaluation.

Market deployment stage(s)

- All.

Barrier(s) addressed

- Renewable technologies are not well-known in the market.
- Renewable technologies are only used by early adopters and in niche markets.
- Laggards are not using well-established renewable technologies.
- Progress towards goals is less than expected.

Description

Once targets are established, the next step is to devise strategies to achieve those targets. These strategies may involve a wide variety of interventions, including stick and carrot programmes, as well as research and information dissemination, and should be backed-up with a solid evaluation framework to assess progress. A good assessment of the technical, economic and achievable potential will provide a basis for the development of a portfolio of programmes, which should include a mix to target different market segments and different barriers.

Additional guidance

The 'classic' approach to market transformation provides one example to follow in devising plans.[1] In the initial stages, the focus is on research, development and demonstration, including experimentation with pilot programmes involving tailored interventions with manufacturers, distributors and other trade allies.

This is followed by efforts to promote early adoption and deployment through strong incentives and through the use of information and marketing efforts including labelling and endorsements. Once the market transformation has begun, codes and standards can be used to capture the rest of the market, while more advanced technologies and practices are promoted and the process begins again.

This 'classic' approach is not necessarily applicable to all situations and, in fact, there is no one-size-fits-all prescription. Another view of a potential portfolio of instruments to be deployed in the various phases of market transformation, similar to that used by Natural Resources Canada in promoting Energy Star high efficiency products, is shown in Exhibit 10. Although aimed at energy efficiency, the instruments are equally applicable to the challenges of promoting renewable heating and cooling technologies.

Examples

The key to Upper Austria's success in renewable energies lies in the integrated Energy Action Plans, which have guided its renewable energy policy over the past 16 years. The Energy Plans of Upper Austria do not focus on a single market barrier or technology, but provide a well-defined mix of support mechanisms, which have led to a strong development in the use of (small-scale) biomass and solar thermal technologies. Some of the mechanisms include:

- Financial incentives.
- Information and public awareness raising.
- Independent advice on energy issues.
- Legal/regulatory measures, including a de facto obligation for the use of renewable heating in new buildings (a requirement in the housing subsidy programme).
- Development of new business models for the sales of renewable energy technologies, especially in the form of contracting models.
- Training of professionals.
- Support of regional R&D and research institutes.

The Energy Plans are largely developed by the *O. Ö. Energiesparverband*, the energy agency of Upper Austria, which is also responsible for most of its implementing measures. The plans are approved by the regional government of Upper Austria. By closely involving the energy agency in charge of most of the implementation, know-how from the previous plans and implementation is fed back into the development of the each new plan.

The Energy Plans try to address the various challenges and barriers through targeted implementation measures. Upper Austria follows a step-wise approach, which aims at extending renewable heating and cooling to more applications,

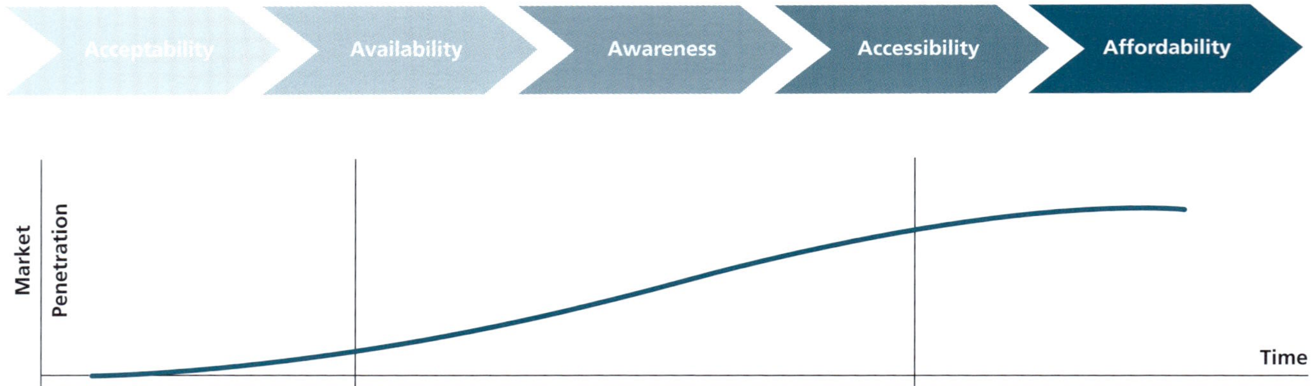

Phase	Initial Deployment	Mass Market	Full Market
Goal	Experiment, prove acceptability, create availability	Capture significant portion of market, encourage availability and affordability	Capture laggards and maintain
Methods	Pilot, give away, direct install	Incentives (with phase out), marketing, training	Regulation, innovative marketing, training field support

Exhibit 10 Market transformation approach
Source: Natural Resources Canada

building types and commercial sectors. While the typical one- and two-family houses are already well-covered by the implementation measures (information, advice, grants) other sectors and market segments need special attention and strategies. For example, larger solar thermal installations in residential building blocks need better trained planners and installers, and the buildings' owners, often large housing companies, are approaching investment decisions differently from the private home owners. In the next phase the *O. Ö. Energiesparverband* aims at specifically targeting industrial companies when promoting solar thermal solution.

With strong political backing and the agreement of longer-term targets in the Energy Plans, Upper Austria has created a stable support framework which has resulted in a positive investment climate. The stop and go support so often seen in other regions and countries was avoided and the renewable heating sector is able to trust that support will not suddenly end in Upper Austria.

The mix of measures used by the state of Upper Austria has proven highly successful in stimulating the demand for renewable heating and cooling solutions. In carrying out the Energy Plans, the regional energy agency relies on three pillars: carrots (such as direct grants for the installation of renewable energy systems), sticks (such as new renewables obligation in new buildings) and guidance (such as independent energy advice given to end consumers).

The implementation addresses the various barriers to growth and aims at a step-wise build-up of a self-intensifying market – both on the demand side and on the supply side. While the former is addressed mostly through information campaigns, independent energy advice and direct grants, the latter is being addressed through the Network of Green Energy Businesses, *Ökoenergie Cluster*, the training of professionals and the support for energy R&D.

Underlying this mix of measures is the experience that the market only grows as fast as its slowest participants (manufacturers, planners, installers, consumers). This approach therefore avoids the sometimes one-sided approaches seen in other programmes, which focus only one or two barriers to growth, but neglect other important growth factors.

10.2 Best Practice 2: Consider having a third party design, implement and/or evaluate the programme

Programme phase(s)

- Portfolio planning (if third party will design programme).
- Programme design (if third party will implement and/or evaluate programme).

Market deployment stage(s)

- All.

Barrier(s) addressed

- Funding or organizing agency does not have the personnel and/or REHC experience to design, implement and/or evaluate a programme.

Description

Many renewable thermal energy support programmes are financed by government bodies. However, many successful programmes have been designed, implemented and/or evaluated by third parties. When the financing body is not experienced with REHC programmes, it is advisable to engage third parties with more experience. These third parties are often regional or national energy agencies, specialized administration offices or private consulting companies.

In addition to lacking experience, government bodies often do not have the personnel capacities necessary to implement programmes. Specialized third parties are adapted to project-based work and are accustomed to temporarily providing additional work force. Finally, the involvement of third parties may help to reduce programme administration costs.

Third parties are particularly appropriate for performing programme evaluations. The independence of evaluators is important for the objective review of programme processes and results and the identification of necessary improvements. Consortia of different organizations, including research institutions, consulting firms and industry associations, can be an additional option for evaluation.

Additional guidance

After the identification of general programme objectives, the financing party (often a government body) should evaluate whether the programme would be best designed, implemented and/or evaluated by the financing party or by a third party. If the financing party is not well-positioned for these tasks, the financing party should identify third parties capable of designing and implementing the necessary actions. In the case of guidance campaigns, private companies experienced with promotion campaigns may be best suited for the work. In some countries, there are specialized agencies such as the Federal Office for Economy and Export Control (BAFA) in Germany who are responsible for subsidy programmes similar to the German *Marktanreizprogramm* (MAP).

Examples

The *Spar mit Solar* programme (Steiermark, Austria) operated within the framework of the Austrian national solar promotion campaign *klima:aktiv*. Because the Steiermark market for solar thermal installations had stagnated in 2003/2004, this programme was designed to stimulate further market growth by overcoming the lack of general knowledge about both the technology and the regional financial incentives provided by the state. The project idea for *Spar mit Solar* was developed by AEE INTEC, a private company with significant experience in designing similar projects and well-established connections within the renewable energy and solar thermal industry sectors. The programme was designed in close cooperation with the relevant (semi-)public bodies such as the energy department of the regional government and the regional energy agency. AEE INTEC was also fully responsible for the implementation of the promotion campaign while programme evaluation was done by the financing parties.

The *Marktanreizprogramm* (Market Stimulation Programme or MAP) is the main instrument of the German federal government to stimulate the uptake of renewable heating and cooling technologies. In its latest edition, the MAP financially supports the installation of solar thermal, biomass heating and heat pump systems. For smaller installations, the incentives are given in form of a grant per square metre of solar thermal collector area, kilowatts of biomass heat capacity or square metres of floor area, in the case of heat pump systems. For large installations, the MAP provides attractive loans (fixed interest rates, no pay-back in the first years and so on). The MAP is under the control of the Federal Ministry of the Environment and is administered by the third parties Federal Office for Economy and Export Control (BAFA) (for the direct grant part of the programme) and the public KfW bank (for the loan part of the MAP). The MAP has been regularly evaluated by several independent scientific institutes (ZSW, Solites, ITW and others).

The Barcelona Solar Thermal Ordinance, implemented by the municipality of Barcelona, requires that new buildings or buildings undergoing major renovations have a solar domestic hot water system that meets at least 60 per cent of hot water demand. The Barcelona Energy Agency is the administrator for this programme. The Energy Agency was started in 1995 to promote renewable energy in the metropolitan area of Barcelona. It began as a project sponsored by the European Union (EU) to encourage cooperation between local politicians, organizations and the public to develop energy projects and was originally called the BarnaGEL. However, in 2000, funding from the EU was replaced by support from the municipality as a non-governmental organization. The agency consortium includes the city council, public institutions and organizations, and two universities. The staff is made up of approximately 20 people.

The Household Subsidy Programme in Norway offered investment support and education measures for the implementation of heat pumps, wood pellet stoves and automated temperature controls. The subsidy covered 20 per cent of all costs including materials and installation, as long as the technology was purchased from a certified dealer and the installation was done by a certified installer. This programme was run by Enova SF, a public agency that exists to design and implement renewable energy projects. It was chosen to run the programme because it had a pre-existing structure with personnel (40–50 people) for such programmes. The Household Subsidy Programme required about 14 staff members, however, its operating budget was tenfold smaller than that of Enova SF's operating budget. Enova SF was also responsible for monitoring the programme and taking corrective actions. Monitoring was based on statistics regarding pre-approval, and so on, and questionnaires collected during the programme implementation.

10.3 Best Practice 3: Break longer-term targets into shorter-term milestone targets

Programme phase(s)

- Portfolio planning.
- Programme implementation.
- Programme evaluation.

Market deployment stage(s)

- All.

Barrier(s) addressed

- Long-term goals appear too hard to meet.
- Long-term goals are not motivating to politicians and the general public, who are focused on shorter-term goals.
- Long-term objectives are not motivating because it is difficult to tell what progress is being made.

Description

Long-term targets are important in setting direction and motivating populations, trade allies and staff. However, specific short-term targets also need to be established to guide and adjust short- and medium-term planning, as well as to measure progress. Short-term targets also assist in determining if corrective action is needed.

Additional guidance

The establishment of milestone or interim targets should be accompanied by the development of an appropriate measurement, verification and reporting system, using accepted standards for measurement and verification. Mechanisms should be established to receive reports and take corrective action if necessary (for example, by increasing incentives or raising the price of conventional fuels) to get progress on track to achieve the next interim target.

Examples

The Climate Alliance of European Cities is a network of European cities and municipalities working at the municipal level on climate protection. These cities and municipalities have entered into a partnership with indigenous rain forest peoples of the Amazon. There are two main objectives of the Climate Alliance: to reduce greenhouse gas emissions to a sustainable level in the northern hemisphere and to support projects to conserve the rain forest and improve quality of life in the southern hemisphere through cooperative programmes.

When the Climate Alliance adopted a new target of 50 per cent reduction in greenhouse gas emissions by 2030 from a baseline year of 1990, they noted that municipal governments are normally only accountable for a short period of time. As a result, it was decided to establish interim targets of reducing greenhouse gas emissions by 10 per cent (based on 1990 levels) every five years.

10.4 Best Practice 4: Design and implement evaluations and refine both portfolio targets and programmes based on the evaluations

Programme phase(s)

- Portfolio planning.
- Programme design.
- Programme evaluation.

Market deployment stage(s)

- All.

Barrier(s) addressed

- It is not clear what is achievable and which programme works best.
- Programme concept has not been tested in target population.

Description

The common saying 'you can't manage what you can't measure' is applicable to any government initiative, and policies for renewable heating and cooling are no exception. Good public policy demands both accountability for the use of public funds and the government's coercive powers, as well as information on what is working and what is not working in order to reassess plans, terminate programmes that don't work and redesign those that could work better. All of this requires the inclusion of appropriate evaluation frameworks.

Additional guidance

An appropriate evaluation framework should include:[2]

- A clear statement of programme goals and a logic chain explaining how inputs and activities, leading to immediate, intermediate and ultimate outcomes achieve those goals.
- Identification of indicators and targets for key results, along with plans for ongoing measurement, verification and reporting.
- Statements of responsibilities for measurement and reporting and accountability for results.
- Identification of key evaluation issues, including evaluation of impacts and cost-effectiveness, assessment of relevance and review of processes.
- Preliminary plans for data measurement, collection and analysis.
- A process for review of evaluation results and decision-making.

Example

The key to Upper Austria's success in renewable energies lies in the integrated Energy Action Plans, which have guided its renewable energy policy over the past 16 years. The implementation of the Upper Austria Energy Plans is monitored regularly. The government publishes an annual implementation report, which is prepared by the *O. Ö. Energiesparverband*. These reports provide detailed data on energy usage, the numbers of renewable energy installations and their energy production. They also report on the measures taken to further decrease the energy demand and to increase the use of renewable energies. However, a critical analysis of individual implementation measures, their effectiveness and possibly even efficiency is largely missing in these reports.

10.5 Best Practice 5: Develop flexible plans and tools to support regional or municipal progress to the established targets

Programme phase(s)

- Portfolio planning.
- Programme design.
- Programme evaluation.

Market deployment stage(s)

- All.

Barrier(s) addressed

- Regional and/or local governments do not have the capacity or expertise to develop and implement programmes.

Description

Sub-national governments, particularly smaller municipalities, do not usually have the capacity and expertise to develop and implement elaborate programmes. If they are to contribute to achieving targets, they need support in the form of training, guidance and flexible tools and templates that they can use as a basis for standardized programmes.

Additional guidance

Associations of municipalities are often effective gatekeepers and clearinghouses for information sources for their members. This means they are well-positioned to inform their members about renewable thermal technologies, existing programmes and new programme development. Senior governments can assist them by funding the development of best practice guides or providing assistance in benchmarking and accessing existing information.

Example

The Climate Alliance of European Cities is a network of European cities and municipalities working at the municipal level on climate protection. These cities and municipalities have entered into a partnership with indigenous rain forest peoples of the Amazon. There are two main objectives of the Climate Alliance: to reduce greenhouse gas emissions to a sustainable level in the northern hemisphere and to support projects to conserve the rain forest and improve quality of life in the southern hemisphere through cooperative programmes.

The Climate Alliance provides municipalities with flexibility as to how they achieve the greenhouse gas emission reductions. Local authorities are encouraged to design and implement suitable local policies and programmes as well as participate in national programmes.

The Climate Alliance has more than 15 different websites to communicate with members and other communities. An online database is being developed of best practices to assist members in the development of effective programmes. The *eClimail* electronic newsletter is distributed to participating members and ministries, providing information about local climate change news, updates from the rain forest and information about upcoming events. In 2008, nine newsletters were sent out by email to 4200 participants and 15 invitations were sent out for events.

More recently, the Climate Alliance has been given a contract to develop a national benchmarking system for local climate protection (Climate Cities

Benchmarking), after the successful completion of the first phase of 'Local Government Climate Partnership'.

10.6 Best Practice 6: Establish clear, consistent, ambitious but achievable portfolio targets

Programme phase(s)

- Portfolio planning.
- Programme design.
- Programme evaluation.

Market deployment stage(s)

- All.

Barrier(s) addressed

- Goals are not credible and do not motivate action.
- Goals are easily met and therefore do not motivate action.
- Goals keep changing.
- Goals are confusing.
- Clear programme objectives, targets and performance indicators have not yet been established.

Description

Long-term visionary strategic goals or objectives are important in setting direction and motivating populations, trade allies and staff. However, they cannot just be 'imagined'. For goals to be credible and truly motivating, they need to be understandable and measurable. They should be based on a solid understanding of the technical, economic and practical potential. They should be challenging (in other words, require extra effort) but achievable (based on realistic assumptions and expectations). Finally, they should be consistent and stable over time.

Additional guidance

For portfolio targets to be understandable and measurable they need three attributes:

1 A clear indicator (including unit of measurement), for example, renewable capacity (MWth).
2 A target date, for example, by 2020.
3 An absolute or relative target number, for example, total of 3000MWth, additional 1000MWth, or 50 per cent increase. If the number is relative,

then the base year must also be specified – for example, 50 per cent increase from 2005.

For consistency and stability, changes to targets should be avoided. In particular, the temptation to restate the base year should be resisted. Similarly, changes in indicators can cause confusion in measurement and reporting. If a target must be changed, it is best to change the target date or target number.

Evaluating potential is key to setting credible yet ambitious targets. This requires studies to evaluate the state of the technologies, the resources and the relative costs compared to conventional energy sources. Forecasting and scenario development are difficult tasks requiring many assumptions and judgments, but there is no substitute.

Examples

The key to Upper Austria's success in renewable energies lies in the integrated Energy Action Plans, which have guided its renewable energy policy over the past 16 years. The government of Upper Austria's 'Energy Future 2030' outlines the path for the next two decades. This programme foresees a 39 per cent reduction of the heat demand by 2030, which, together with the increased use of renewable energies, will lead to 100 per cent of the heating demand being covered by renewable energy sources. The goal is clear and the target is measurable and long-term. Although this goal was established in 2007, it is an extension of goals that have been stable and consistent since 1993. The target is precise and its achievability, while not assured, is informed by the experience of the previous 16 years.

The Climate Alliance of European Cities is a network of European cities and municipalities working at the municipal level on climate protection. These cities and municipalities have entered into a partnership with indigenous rain forest peoples of the Amazon. There are two main objectives of the Climate Alliance: to reduce greenhouse gas emissions to a sustainable level in the northern hemisphere and to support projects to conserve the rain forest and improve quality of life in the southern hemisphere through cooperative programmes. The original goal of the Climate Alliance was to reduce greenhouse gas emissions by 50 per cent by 2010. This was a visionary goal but it was not based on a real analysis of the potential. In 2005, the Climate Alliance prepared actual estimates of emission reductions in Munich, to reassess the goal. As a result, in 2006, a new goal of 50 per cent reduction from 1990 levels by 2030 (with interim targets) was adopted. The goal is clear and the target is measurable and long-term. It is unfortunate that it had to be changed (resulting in loss of momentum, stability and credibility), but the fact that it is now grounded in a credible understanding of the potential means that future changes are less likely and there can be confidence in the plan.

10.7 Best Practice 7: Identify and encourage 'early adopters'

Programme phase(s)

- Programme design.

Market deployment stage(s)

- Initial deployment.

Barrier(s) addressed

- Specific technology or application is not well known.

Description

When new technologies or programmes of any type are introduced, some individuals or organizations move quickly to adopt these technologies and programmes. These parties are often called 'early adopters'.

Early adopters of renewable energy technologies are motivated by a variety of reasons such as a desire to reduce their environmental impact, interest in the technology or long-term financial savings. Identifying and supporting these individuals or organizations can be a way to facilitate uptake.

Additional guidance

Potential early adopters should be identified and selected using criteria such as openness to innovation and potential multiplication effects. Market studies should be done to determine what specific measures should be taken to encourage early adopters. Information about their participation should be broadly disseminated.

This best practice is particularly applicable in situations where the usefulness of a programme depends on the participation of the majority of households or organizations concerned. This includes not only industrial certification schemes but also general market uptake of new technologies.

Examples

The Solar Keymark is a certification mark owned by the European Committee for Standardization (CEN). It certifies conformity of a product with the applicable European Standards: EN 12975 for solar thermal collectors or EN 12976 for factory-made solar thermal systems. In the first years of the programme, a chicken-egg problem existed: most solar thermal companies did not see a good reason to pay for the Solar Keymark certification as there were almost no support programmes which linked eligibility to this certificate. Public authorities in charge of solar thermal support programmes, on the other hand, did not take the Solar Keymark seriously as there were almost no Solar Keymark certified products out in the market. Not until 2006 did the Solar Keymark really take off, and now there are more than 700 Solar Keymark certified products available on the market. An important factor for the success of the Solar Keymark programme was the participation of 'early-adopters' from the industry and the national governments, who realized the potential advantage of having such a common certification scheme.

10.8 Best Practice 8: Consider offering additional incentive amounts for higher quality equipment or additional efficiency measures

Programme phase(s)

- Programme design.
- Programme implementation.

Market deployment stage(s)

- Mass market.

Barrier(s) addressed

- Current REHC installations are of poor quality. For example, inexpensive solar hot water systems are being installed.

Description

In order to further improve quality and efficiency of the supported technology, incentives can contain bonuses or other mechanisms to create an additional pull for quality products. The higher incentive can be granted based on the achievement of a certain quality certificate or label.

A higher incentive could also be offered for other energy efficiency measures implemented at the same time, such as the replacement of an old gas boiler with a new condensing gas boiler when a solar thermal system is installed.

Additional guidance

Depending on the technology, suitable requirements should be identified, such as quality certificates or labels, or higher certified performance values. Where possible, existing quality standards based on established (international) standards should be referenced. Where the installation of a new renewable heating system could be reasonably combined with other improvements in the energy efficiency of the buildings, such combined improvements could receive additional incentives.

The additional incentive should not lead to windfall profits, in other words, it should not support a quality level or other energy efficiency measure that would normally be implemented anyway. The goal is to encourage better-than-normal energy performance.

Examples

The Salzburg Regional Subsidy Programme (Austria) is designed and implemented by the regional government administration. The regional government finances the subsidies and decides annually on the budget for the next year. In 2009, the total amount of the subsidy is limited to a maximum of 30 per cent of the total investment. The programme provides a base incentive for REHC equipment, which can then be increased by meeting additional requirements. In the case of a wood pellet boiler, the 2009 base incentive is fixed at €1000. The total incentive amount can be increased to up to 30 per cent of the total investment by:

- Installing a wood pellet boiler which meets the high efficiency requirements (€400).
- Installing buffer storage tanks (€500).
- Performing a hydraulic adjustment (€200).
- Installing high efficiency pumps (€50 per pump).
- Providing particle removal (€500).
- Performing an energy audit according to specific guidelines for biomass heating (€200).
- Performing a heating system inspection according to clear guidelines (€100).
- Insulating the building to a certain level (€100–1000).
- Installing a heat recovery system (€30–500).
- Combining the boiler with a solar thermal system (€500).
- Changing the heating system from fossil to biomass fuel (€500).

These bonuses make the investment in additional efficiency measures more attractive to the end-consumer. Most applicants ask their installers to plan the installation in a way that meets these efficiency requirements. As a result, most of the installations subsidized are state-of-the-art.

10.9 Best Practice 9: Consider requesting technical drawings from installers

Programme phase(s)

- Programme design.
- During programme implementation, if poor installation quality is a problem.

Market deployment stage(s)

- Mass market.

Barrier(s) addressed

- Installation issues such as oversized systems or plumbing mistakes are a problem. For example, biomass boilers may be over-dimensioned for the heat load of the building.

Description

Poorly installed systems discourage market uptake because they are more expensive or less effective than properly designed and installed systems. In order to keep costs down and to generally increase the quality and efficiency of subsidized technologies, programmes should include measures for controlling quality. One possible way to prevent basic mistakes is to request technical drawings of the installation during the application process, prior to installation.

Additional guidance

In general, the application and approval process of a subsidy programme should be kept as simple as possible in order to minimize programme costs and to make the programme attractive to potential users. However, the programme evaluation should include random on-site inspections in order to monitor the quality of subsidized installations. If basic mistakes are found on a significant scale, measures to improve installation quality should be considered and programme directives should be adapted accordingly. A simple and cost-efficient way to avoid basic installation mistakes is to request technical drawings from the installer before installation work starts.

Since basic mistakes are most likely to occur in areas without a trained and experienced installation industry, measures for quality control will be most useful in these areas.

Examples

The Salzburg Regional Subsidy Programme (Austria) is designed and implemented by the regional government administration. The regional government finances the subsidies and decides annually on the budget for the next year. In 2009, the total amount of the subsidy is limited to maximum of 30 per cent of the total investment. For the installation of a pellet boiler, for example, €1000 can be paid as a base sum. Additional subsidies are paid for increasing the overall installation efficiency, for example, by using highly efficient pumps. The programme administrators found that the subsidized installations often had significant flaws. The programme was modified to require that the installer turn in a plan for the installation to allow for checking major quality issues before the installation is approved. To simplify the application process as much as possible, the processing of applications was automated and a dedicated webpage was established, presenting information on the subsidy scheme and allowing for online application.[3] All steps during the application process are automated and can be carried out online, including handing-in of technical drawings. Technical drawings are then checked by external experts and compared to the energy certificate (also by the subsidy programme) so that mistakes such as oversizing the heating systems are avoided.

10.10 Best Practice 10: Regulate the quality of the equipment and installations funded with incentives

Programme phase(s)

- Programme design.
- Programme installation.

Market deployment stage(s)

- Mass market.

Barrier(s) addressed

- Equipment currently being installed is not of good quality.

Description

While most purchasers are usually interested in a good quality renewable heating system, they sometimes buy low-quality products that are not as efficient or durable as their higher-quality counterparts. Furthermore, the planning and installation of the system has a strong influence on the overall working of the heating system.

Support programmes should therefore include suitable quality criteria to avoid supporting poorly-performing products and installations. This is important not only for accountability reasons but also because happy users of renewable installations are the best marketing tool for the development of these markets.

Additional guidance

Where possible, existing quality standards based on established (international) standards should be referenced. This allows the supplying manufacturer to have products tested/certified only once.

The criteria should be ambitious enough to have an effect on the market (for example, increase of the average performance of systems installed), but must take into account the state of the market infrastructure. Overly harsh requirements can be counterproductive and prevent market growth.

In awareness-raising campaigns, homeowners should be informed about quality differences and methods of identifying good quality products and services.

Examples

The Household Subsidy Programme in Norway offered investment support and education measures for the implementation of heat pumps, wood pellet stoves and automated temperature controls. The programme successfully increased the demand for heat pump systems and, to a much lesser degree, of wood pellet stoves. Because experiences in the past had shown that poor quality systems (hardware and/or installation) had reduced the efficiency of these technologies and hurt their positive image, the government decided to set strict quality criteria in its financial incentive programme. Only products sold by approved dealers and installed by approved installers were eligible for the support. This programme ran only for a few weeks but supported approximately 20,000 installations. No problems with the quality of these installations were reported.

The Barcelona Solar Thermal Ordinance, implemented by the municipality of Barcelona, requires that new buildings or buildings undergoing major renovations must have a solar domestic hot water system that meets at least 60 per cent of hot water demand. The ordinance has established clear requirements for the obligatory solar domestic hot water systems in new buildings and buildings undergoing major renovation. Installation plans must be provided in the building planning phase and occasionally the Barcelona Energy Agency sends an inspection team to the building site, to ensure that certified equipment is properly installed and that the solar domestic hot water heater is in good working order. Strict quality requirements are necessary to ensure a good image of the solar thermal technology, which had been tarnished by poor installations during the 1980s.

The French tax credit programme was implemented as a direct fiscal measure for the support of renewable heating. The tax credit applies to household solar domestic hot water and other solar heating, biomass heating, and heat pumps. The programme for solar thermal installations also includes clear quality requirements, and the collectors have to be certified either with the Solar Keymark or with the French CSTBat.

10.11 Best Practice 11: Design and implement long-term programmes

Programme phase(s)

- Programme design.

Market deployment stage(s)

- Mass market.
- Full market.

Barrier(s) addressed

- Short-term programmes have not produced the desired results.

Description

Programmes with long-term perspective and stability are more likely to achieve sustained market growth. Stability of framework conditions is an important parameter for investment decisions of the supply side. Support programmes should, in most cases, aim for long-term programmes that create a stable and positive investment climate. Stop and go support often leads to unhealthy dynamics in the market (sudden growth and sudden contractions) and to insecurity; creating new barriers in the market. Also, many programmes require administrative personnel be hired or trained. In long-term programmes these start-up costs play only a minor role.

Additional guidance

Ensuring the long-term existence and stability of a support programme is easiest to achieve when the effect on the public budget is small. Otherwise, programmes often face annual budget approvals with uncertain results. Many support measures need no or only little public finances. For example, a building code requiring the installation of renewable heating systems in new buildings involves some administrative overhead, but is otherwise budget neutral. Similarly, consumer information requirements such as energy performance certificates or energy labels can be used to promote renewable heating without the need for significant public financing. Some innovative financing schemes provide incentives for renewable heating largely outside the public budget. An example is the renewable energy certificate system in Australia which allows solar hot water system owners to sell their certificates to organizations required to meet renewable energy targets.[4] Where public budgets must be used to finance the programme, sufficient funds should be ensured and the programme should be designed to run for several years.

This best practice is generally applicable to all support programmes. However, the 'harder' the measure, the more important stability is over time. For example, a building code requiring the installation of renewable heating equipment should not change very often and any changes should be made known well in advance to allow the construction and renewable heating equipment industries to plan ahead.

Examples

The *Marktanreizprogramm* (Market Stimulation Programme or MAP) is the main instrument of the German federal government to stimulate the uptake of renewable heating and cooling technologies. In its latest edition, the MAP financially supports the installation of solar thermal, biomass heating and heat pump systems. For smaller installations, the incentives are given in form of a grant per square metre of solar thermal collector area, kilowatts of biomass heat capacity or square metres of floor area, in the case of heat pump systems. For large installations, the MAP provides attractive loans (fixed interest rates, no pay-back in the first years, and so on). The MAP has been running almost continuously since 1999. The grant approval has only been stopped a few times due to the depletion of the annual budget. Several times this stop was avoided by significantly lowering the specific incentive offered. The long-term approach of the MAP has led to a high awareness, acceptance and efficiency among all involved stakeholders: the environment ministry, the authority administering the grants (BAFA), the suppliers of REHC equipment, the planners and installers as well as the consumers. The ten years of positive experience has led to a certain confidence in the market that the programme will be available in the future, thus creating a positive market framework in which companies invest in the expansion of the REHC business. Homeowners do not have to fear that the programme will be unavailable if they do not buy immediately, establishing a stable and increasing market for REHC products. The relative stability of the programme should not conceal the fact that the specifics of the MAP have changed over time. The MAP has been modified ten times since its inception. The modifications changed the level of support for the different technologies and applications, as well as the requirements for eligibility under the programme. The modifications also included changes in the application procedures and processing. Nevertheless, the changes were small enough to allow for solid capacity-building for all involved stakeholders, including the grant authority (BAFA). The changes were discussed in advance with the relative industries, which helped increase acceptance of the programme and its implementation. The government and the industry agree that the stability of the MAP has been one of the success factors of this REHC support programme.

The energy strategy for the residential sector in Vorarlberg (Austria) includes two types of financial incentives provided to households for the installation of REHC technologies: the housing programme and a direct subsidy programme. Both are state designed and financed. The housing programme provides low-interest loans for building or renovating houses and the installation of district heating (biomass derived) or highly efficient gas, biomass or heat pump technologies. Except for district heating systems, the main heating system has to be combined with a solar thermal energy system. The state of Vorarlberg provides additional direct subsidies for heat pumps, biomass heating and solar thermal installations. The Vorarlberg Regional Subsidy programme has provided financial incentives for solar thermal since 1991 and today solar thermal is installed in the majority of new residential buildings. With the help of this programme, heat pumps have also become well-established in the market and they are installed into 63 per cent of new buildings. While most of the installations today are due to the requirements in the housing programme, the additional subsidy motivates consumers to choose systems with higher quality and efficiency. End-consumers demand that installers meet the strict technical requirements of the regional subsidy programme.

10.12 Best Practice 12: Offer incentives by energy (kWh) or capacity (kW or m²) rather than as a percentage of cost

Programme phase(s)

- Programme design.
- Programme implementation.

Market deployment stage(s)

- Mass market.

Barrier(s) addressed

- Programme includes incentives that have the potential to distort the market in a negative direction.

Description

Financial incentives are always tied to a certain parameter, such as capacity or first cost. In order to not encourage price increases, incentives should not be related to the costs themselves. Ideally the offered incentive is linked to energy-related parameters, such as estimated (or measured) kilowatt hours of heat produced from renewable energy, or kilowatts of installed capacity.

Where non-energy related parameters are used (such as 'per installation' or 'per square metre of solar thermal collector area'), other requirements should be added to the financial incentive programme, for example, a minimum performance per square metre of collector area, or a minimum contribution to the overall heat demand of the building.

Additional guidance

This best practice could be implemented in two steps:

Identification of suitable parameters to which the financial incentive can be linked

As the support programme aims at reducing conventional energy consumption, the incentive should ideally be linked to the expected savings of conventional energy. However, it is often not possible or practical to assess the savings for each individual installation. Therefore other parameters, such as the measured or expected production of heat from renewable energy or the installed capacity should be used.

The parameter should be as simple as possible to assess in order to avoid additional costs for both the administration and the homeowner. For example, the installed capacity of a solar thermal collector may be available from a test report from an independent test institute.

Wherever possible, performance parameters should be based on accepted international standards. This reduces manufacturer testing costs and encourages trade.

10.12

Identification of suitable level of the incentive

Identification of suitable level of the incentive may be the most subjective step in the design of the incentive programme. In some cases, the incentive is necessary to balance the lower market costs of a comparable conventional installation, in others the renewable energy technology may be almost cost competitive and the incentive serves mainly as a psychological incentive signalling the government's endorsement of this technology. The 'right' level can be fine-tuned over time by periodic adjustments.

Examples

The *Marktanreizprogramm* (Market Stimulation Programme or MAP) is the main instrument of the German federal government to stimulate the uptake of renewable heating and cooling technologies. In its latest edition, the MAP financially supports the installation of solar thermal, biomass heating and heat pump systems. The MAP bases its incentives on the square metres of collector area in the case of solar thermal, the kilowatts of capacity in the case of wood pellet boilers, and the square metres of floor area in the case of heat pump systems.

The Austrian state of Steiermark provides direct subsidies for the installation of efficient biomass heating and solar thermal collectors through the *Umweltlandesfonds*. For solar thermal collectors, this programme provides a subsidy of a fixed amount plus an additional €50/m^2 of collector.

A negative example can be found in the French tax credit programme implemented as a direct fiscal measure for the support of renewable heating. The tax credit applies to household solar domestic hot water and other solar heating, biomass heating and heat pumps. Since installation costs were not eligible for the credit, certain installers were providing 'free' installation for overpriced equipment. This artificially increased the amount of credit provided. Also, solar thermal collectors have increased in price by 7 per cent annually in conjunction with market growth. An increase in raw material prices has been given as a reason, however, it is possible that industry players are taking advantage of the tax credit rather than passing these savings to the customer.

10.13 Best Practice 13: Penalize non-compliance with 'stick' programmes

Programme phase(s)

- Programme design.
- Programme implementation.

Market deployment stage(s)

- Full market.

Barrier(s) addressed

- Programme includes regulations which may be ignored or are not being followed, fully.

Description

When a programme introduces mandatory requirements, such as the installation of renewable heating installations in new buildings, it must also plan negative consequences for those who do not comply with the regulation or law. If not, the programme will be seen as a voluntary one and not attain the expected results.

Additional guidance

This best practice has two implementation issues:

Check mechanisms

The 'stick' programme must contain check mechanisms, which allow the identification of non-compliance. Where checks of each installation/building are impossible or impractical, random checks may be made.

Where suitable, checks by the programme authority can be replaced by awareness in the market itself: where the non-compliance is easy to spot, citizens or, in the case of commercial homeowners, competing companies, can be encouraged to point out non-compliance.

Consequences

The consequences for non-compliance must be chosen in relation to the seriousness of the non-compliance. For example, where a renewable heating system is required in new buildings, the failure to install such a system at the time of construction is a serious violation of the law or regulation and should be punished more harshly than the disregard of a minor quality or efficiency requirement. The consequence should involve a financial punishment that is high enough to discourage others from following the bad example. Where reasonable, the consequence should also involve the rectification of the non-compliance.

This best practice applies to both the design and implementation phase: during the design phase, a negative consequence and a mechanism to enforce it must be planned into the programme. During the implementation phase, the consequences must be carried out in order to 'punish' those that do not comply with the requirements.

Examples

The Barcelona Solar Thermal Ordinance, implemented by the municipality of Barcelona, requires that new buildings or buildings undergoing major renovations must have a solar domestic hot water system that meets at least 60 per cent of hot water demand. The ordinance is enforced by random installation checks and fines of €6000–60,000 for the violation of the ordinance. During the building planning phase, the homeowner must provide an installation plan, which is checked and, where necessary, improved by the Barcelona Energy Agency. Occasionally the agency sends an inspection team to the building site, to ensure that certified equipment is properly installed and that the solar domestic hot water heater is in good working order. The checks and fines ensure that installers and home owners fully respect the ordinance.

10.14 Best Practice 14: Develop and run programme in close cooperation with relevant public bodies and industry partners or organizations

Programme phase(s)

- Programme design.

Market deployment stage(s)

- All.

Barrier(s) addressed

- Funding or organizing agency does not have the personnel and/or REHC experience to design, implement and/or evaluate a programme.

- REHC proponents and industry members are not yet working together. For example, energy agencies and industry members may not yet be involved in cooperative activities.
- Legal or administrative barriers will hamper programme. For example, building codes requiring expensive or inconvenient inspections of solar installations might discourage installation, or lack of widely-accepted testing standards might limit access to markets in different regions.
- Other REHC or energy conservation programmes are operating in the target geographical area.

Description

When designing and implementing a support programme, relevant stakeholders, such as related government agencies and non-governmental stakeholders, should be involved. Many problems with support programmes can be avoided through a participatory process which ensures that important parameters are not overlooked. Good cooperation can be achieved both through formal processes, such as public hearings, the setting-up of steering committees with external stakeholders and liaisons with regulatory and standards bodies; and through informal consultation, such as those with industry associations.

Often the affected stakeholders (or their organizations) can point out ways to achieve better results with the same effort, or the same result with fewer unwanted side effects (for example, administrative overhead). Finally, cooperation in itself can help increase acceptance of a support programme by market players.

Additional guidance

The implementation of this best practice could best be achieved taking the following sequential steps:

Identify the relevant stakeholders

Everyone who is or could be affected by or involved in the implementation of the support programme is a stakeholder. For example, the stakeholders for a solar thermal grant scheme include the users/buyers of solar thermal systems, installers, manufacturers, retailers and private banks. Further interest groups, such as environmental non-governmental organizations or local United Nations Agenda 21 groups, may be able to contribute insight. Other governmental bodies could also be affected, such as financial authorities or the department or agency responsible for spatial planning.

Determine desired level of cooperation

The desirable level of cooperation depends on the nature of the support programme. A major change to a piece of legislation, affecting the business model of many companies will need a much closer cooperation than a change in an awareness-raising campaign.

Plan method of cooperation

The concrete way of how to cooperate with the relevant stakeholders depends on many factors. For example, in a small municipality, it may be suitable to invite the affected organizations and businesses directly. In larger municipalities and at the regional and national levels, this is frequently impractical and the professional representations should be contacted instead, for example,the associations of homeowners, installers, architects and industry.

Examples

The *Spar mit Solar* programme (Steiermark, Austria) operated within the framework of the Austrian national promotion campaign *klima:aktiv*. Because the market for solar thermal installations had stagnated in 2003/2004 in Steiermark, this programme was designed to initiate further market growth by providing information on both the technology and the existing regional financial incentives provided by the state. The programme was implemented in close cooperation with a regional newspaper group, which advertised the programme and the information campaign through their subregional newspapers. This media partnership was important for the awareness raising effort and for attracting participants to the eight to ten information events held each year. In addition, companies were invited to establish business contacts after each information event, allowing the increased awareness to be immediately turned into business contacts, which then often led to new installations of solar thermal systems. Without the cooperation of local solar companies, much of the positive effect of the information events would have been lost.

The Solar Keymark is a certification mark owned by the European Committee for Standardization (CEN). It certifies conformity of a product with the applicable European Standards, EN 12975 for solar thermal collectors or EN 12976 for factory-made solar thermal systems. The Solar Keymark is the solar thermal specific edition of the general Keymark certification scheme, which is available for a wide variety of products and services. This scheme was developed by the solar thermal industry in close cooperation with CEN and with support from the European Commission. This cooperation allowed not only the joint development of the specific scheme rules, but also furthered the promotion of the Solar Keymark. Acceptance in national and regional support programmes was vital for the success of the Solar Keymark.

10.15 Best Practice 15: Establish market needs/barriers prior to programme design

Programme phase(s)

- Programme design.

Market deployment stage(s)

- All.

Barrier(s) addressed

- Market needs and barriers have not yet been studied.
- Need to explicitly formulate which household behaviour to address.

Description

Renewable heating technologies typically face specific market needs and barriers. An effective support programme combines several individual measures addressing several different needs and barriers; together they can multiply the effectiveness of each of the measures alone. Before designing a programme, the specific needs and barriers for the specific technology and market segment must be identified.

Additional guidance

There are two general complementary methods of gathering the information necessary to assess market barriers/needs: literature research and surveys.

- **Literature research.** Many of the challenges of renewable heating technologies are well-documented and relevant in many places. Studies by various energy agencies, universities and other interest groups have analysed the specific barriers and needs of different renewable energy technologies. Most of them are publicly available and can be used in the identification of barriers and needs.
- **Surveys**. Other barriers may be specific to the local conditions and can be assessed only locally. A survey of stakeholders can help clarify the market barriers and needs. For example, when homeowners are surveyed on why they did or did not purchase a renewable heating solution, one receives information on the needs of the demand side. If the answer is, 'I did not know this existed, or was useful in our municipality', then such information should be made part of the support programme. If, on the other hand, the answer is, 'I asked my installer, but the costs were so high that we chose a conventional boiler', a suitable programme may involve financial incentives,

lowering the costs of the renewable heating solution. Similarly a survey on the supply side (installers, manufacturers) can help determine their biggest challenges and be used in the design of the support programme.

Examples

The Solar Keymark is a certification mark owned by the European Committee for Standardization (CEN). It certifies conformity of a product with the applicable European Standards: EN 12975 for solar thermal collectors or EN 12976 for factory-made solar thermal systems. The Solar Keymark is the solar thermal specific edition of the general Keymark certification scheme, which is available for a wide variety of products and services. The Solar Keymark certification scheme was a direct result of the European solar thermal industry's identification of multiple testing/certification requirements as a key barrier to an open market for solar thermal products. In the 1990s, manufacturers and importers were confronted with differing national requirements for support programmes: a product eligible in one country was often not eligible in another country or needed to undergo additional testing and sometimes certification in local testing institutes and according to local regulations. This created additional costs and caused delays to entry into new markets. Today, most existing national support programmes accept Solar Keymarked products as eligible in their financial incentive schemes or solar obligations. This has considerably reduced the costs to companies active in several European countries, and reduced the time-to-market for new products.

The *Spar mit Solar* programme (Steiermark, Austria) operated within the framework of the Austrian national promotion campaign *klima:aktiv*. The campaign was initiated because the government had established that a lack of information on the part of the consumers was the main barrier for the broad deployment of solar thermal technology. The campaign focused on raising awareness of and providing information on solar thermal technologies. The campaign included intensive media relations and between eight and ten information events per year. The increase in demand for solar thermal seems to be a direct result of the programme, validating the original hypothesis that a lack of information was the key barrier in the market.

10.16 Best Practice 16: Establish clear programme objectives, targets and performance indicators

Programme phase(s)

- Programme design.

Market deployment stage(s)

- All.

Barrier(s) addressed

- Clear programme objectives, targets and performance indicators have not yet been established.

Description

The identification of a clear objective is an important step in programme design; the objective provides guidance in determining the measures and programme types to be implemented. Most of the successful programmes were designed based upon clearly identified objectives.

Many of the successful programmes analysed have also set specific measurable programme targets, based upon a preparatory market analysis. However, not all market developments can be foreseen (for example, fuel price changes) and sometimes long-term targets can turn out to be too high or too low. Still, clear programme targets or performance indicators facilitate programme evaluation and decisions about corrective measures.

Additional guidance

Setting programme targets must be based on market surveys or on general knowledge about REHC markets and their potentials. By defining business-as-usual scenarios and alternative scenarios, the amount of effort and budget necessary to reach a certain scenario can be determined.

Typically, a measurable target for support programmes is to increase the installed capacity of a given technology by a specific amount. More specific targets can be to improve the average quality or efficiency of heating installations, to reduce greenhouse gas emissions, to create a certain number of new jobs and so on. For guidance programmes, typical targets can be to attract a certain number of visitors to information events, to obtain a certain number of hits on websites or to distribute a certain number of printed brochures, leaflets, and so on.

Whatever the target is, it should be used for programme evaluation by comparing the target set to the achieved results. If deviations can be explained by changing conditions, targets should be adapted for the next programme period. If this is not the case, deviations imply the need for programme revision.

Examples

The *Spar mit Solar* programme (Steiermark, Austria) operated within the framework of the Austrian national promotion campaign *klima:aktiv*. Because the market for solar thermal installations had stagnated in 2003/2004 in Steiermark, this programme was designed to stimulate further market growth by providing information on both the technology and the existing regional financial incentives provided by the state. The programme's primary criterion for success was the initiation of additional uptake of solar thermal systems measured in square metres installed. The target set was to end the stagnation on the solar thermal market and to eventually increase the installed collector area by 10 per cent per year, meaning a total impact of 30 per cent market growth during the project duration. This target was more than achieved: from 2005 on, the market grew by 30 per cent each year, meaning that the solar thermal market in Steiermark almost doubled during the programme duration.

The local regional energy agency in Tirol (*Energie Tirol,* Austria) started the *Ja zu Solar* information campaign in 2004. Since a lack of information on the consumer side was identified as one of the main barriers to increased uptake of solar thermal technologies, the implementation of a broad information campaign was seen as an appropriate tool to reach this goal. The programme's primary criterion for success was the initiation of additional uptake of solar thermal technologies measured in additional square metres installed. For the financing industrial partners, the number of additional business opportunities resulting from their presence at events was another way to measure success. The target set was to increase the installed collector area from 160,000m^2 (2005) to 300,000m^2 within five years. In 2008, a total of more than 400,000m^2 had been installed in Tirol. The boom of the market in 2006 and 2007 was unexpected so that the target set seems too low today.

The Salzburg Regional Subsidy Programme (Austria) is designed and implemented by the regional government administration. The regional government finances the subsidies and decides annually on the budget for the next year. In 2009, the total amount of the subsidy is limited to 30 per cent of the total investment. The programme provides a base incentive for REHC equipment, which can then be increased by meeting additional requirements. The programme's objective is to promote the installation of highly efficient technologies. As a result, success is not measured in terms of market growth achieved. A unique feature of this programme is a detailed catalogue of technical specifications that have to be met in order to receive the subsidy. It can be assumed that the quality of installations in Salzburg is very high compared to other regions, however, no data on this is available.

The Barcelona Solar Thermal Ordinance, implemented by the municipality of Barcelona, requires that new buildings or buildings undergoing major renovations must have a solar domestic hot water system that meets at least 60 per cent of hot water demand. The objective was to require that most new buildings (or buildings undergoing major renovations) meet at least 60 per cent of the domestic hot water load with solar. An overall target of 100,000m^2 of solar thermal collector surface area was set for 2010, of which 40 per cent had been achieved by 2006.

10.17 Best Practice 17: Consider requesting an energy audit certificate with the application

Programme phase(s)

- Programme design.
- Programme implementation.

Market deployment stage(s)

- All.

Barrier(s) addressed

- Other (non-REHC) measures are not being installed, even when they are economically attractive. These may be energy efficiency measures such as additional insulation.
- Installation issues such as oversized systems are a problem. The installation of oversized systems is a widespread problem because installers and equipment sellers want to avoid undersizing problems and because their profits typically increase with increasing system size.

Description

European Community initiatives on climate change and security of supply include the Directive 2002/91/EC[5] on the energy performance of buildings. The directive includes a methodology for calculating the integrated energy performance of buildings and minimum standards on energy performance of building, and systems for the energy certification of new and existing buildings. The main goal of this action is to significantly decrease energy consumption for building-related services. There are comparable systems in other countries. Natural Resources Canada (NRCan) developed the EnerGuide rating service[6] that is provided to owners of new houses by NRCan-affiliated energy advisors. The service includes the analysis of house plans, recommendations for energy-saving upgrades and the verification of applied energy upgrades (including a blower door test) after the construction is done. Finally, the homeowner is provided with an official label for his building. A similar service provides energy audits for existing homes under the ecoENERGY Retrofit programme.[7]

The uptake of energy certification of buildings can be supported by requesting energy certificates with the application for financial incentive programmes. Moreover, because smaller REHC installations may be required when energy efficiency measures are taken first, REHC costs may be reduced. Requesting energy certification before the start of an installation can help to avoid such basic mistakes such as oversizing biomass boilers.

Additional guidance

Consideration should be given to two possible difficulties: the possibility that requesting an energy certificate may present an additional barrier for the homeowner, and the possibility that obtaining an energy audit certificate will encourage the homeowner to choose to implement energy efficiency options instead of renewable heating options. (Some energy audit certificates provide information on payback periods for a variety of energy efficiency and renewable energy measures. The renewable heating/cooling payback periods are not always the most attractive.)

This measure is only applicable where there is an existing energy audit certificate infrastructure. Since basic mistakes such as system oversizing are most likely to occur in areas without a trained and experienced installation industry, measures for quality control will also be most useful in these areas.

Examples

The Salzburg Regional Subsidy Programme (Austria) is designed and implemented by the regional government administration. The regional government finances the subsidies and decides annually on the budget for the next year. In 2009, the total amount of the subsidy was limited to a maximum of 30 per cent of the total investment. For the installation of a pellet boiler, for example, €1000 can be paid as a base sum. Additional subsidies are paid for increasing the overall installation efficiency, for example, by using highly efficient pumps. The programme originally provided a subsidy after the applicant turned in the invoice for the installation. The administration, however, often found the subsidized installations had significant flaws. Now, the installer must submit a plan for the installation that allows for checking major quality issues before the installation is approved. The plan is checked against the building's energy certificate, another prerequisite for eligibility.

10.18 Best Practice 18: Design and implement multifaceted programmes

Programme phase(s)

- Programme design.
- Programme evaluation, if the evaluation shows that programme aspects should be added or removed from the programme.

Market deployment stage(s)

- All

Barrier(s) addressed

- Single-faceted programmes have not produced the desired results.

Description

The development of mature technology markets is usually hampered by a number of barriers. While high investment costs for advanced renewable heating technologies are currently the main barrier in most regions, there are often many additional inhibiting factors, such as lack of public awareness or poorly developed retail and installation networks. Support programmes are more likely to be successful when all of the identified barriers are addressed. In addition, the combination of different programme types should be considered. For example, the impact of 'carrot' programmes can be increased by combining them with guidance programmes. When appropriate, the addition of 'stick' programmes, such as housing regulations, will further increase the impact.

Multifaceted programmes are especially appropriate when several technologies are targeted. Different technologies may face different barriers and therefore require different kinds of support.

Additional guidance

In order to identify the different barriers to be addressed by a multifaceted programme, a market study should be conducted at the beginning of the design process. The following information should be collected:

- **Market status:** Who are the manufacturers? Wholesalers? Retailers? Installers? Operations and maintenance providers? End consumers? Can the existing market infrastructure respond quickly to an increase in demand? (For energy-efficient biomass technologies, both the equipment and fuel supply chain should be included in the market status study.)
- **Market needs and barriers:** What are the needs and barriers, both perceived and real, identified by all of the market players?
- **Existing programmes:** What existing programmes are operating in the region? How are these programmes addressing the identified needs and barriers?

The information gathered during this study should allow programme designers to identify currently unaddressed needs and barriers which can then be targeted by a multifaceted programme. If certain barriers are already addressed by other organizations or programmes, the new programme should be designed to complement not conflict, with existing programmes (See *Best Practice 29: Design programme to complement, not conflict, with other programmes in the region.*)

Examples

The local regional energy agency in Tirol (*Energie Tirol*, Austria) started the *Ja zu Solar* information campaign in 2004. Since a lack of information on the consumer side was identified as one of the main barriers to increased uptake of solar thermal technologies, the implementation of a broad information campaign was seen as an appropriate tool to reach this goal. Around 15 information events with around 250 visitors each were organized annually. The information campaign was launched together with a special direct subsidy for solar thermal installations. The goal of the campaign was not only to spread information on the technological and economical viability of solar thermal technologies but also to inform potential users about opportunities provided by the new subsidies available for solar thermal installations. This programme combined 'carrot' and 'guidance' components.

The *Marktanreizprogramm* (Market Stimulation Programme or MAP) is the main instrument of the German federal government to stimulate the uptake of renewable heating and cooling technologies. The MAP currently financially supports the installation of solar thermal systems, biomass heating installations and heat pump systems. Because financial incentives can only provide one piece of the market stimulation puzzle, the MAP was complemented with awareness-raising campaigns (*Solar – na klar!*, 1999–2002; *Solarwärme Plus* since 2002, *Wärme von der Sonne* since 2005 and *Woche der Sonne* since 2007). Furthermore, to promote technological research and development, programmes such as *Solarthermie2000* and *Solarthermie2000plus* provided financial support for pilot and demonstration projects. Since 2009, the MAP has been combined with the Renewable Heating Law that requires the use of renewable energies to cover a certain minimum percentage of the total energy demand of a new building. The new regulations significantly lower incentives in new buildings because incentives will only be provided when the installation provides a higher percentage of the energy demand than what is legally required. This is an example of a programme combining 'carrot', 'guidance' and 'stick' components.

The French tax credit programme was implemented as a direct fiscal measure for the support of renewable heating. The tax credit applies to household solar domestic hot water and other solar heating, biomass heating, and heat pumps. For 2009, the credit for biomass heating has been reduced to 40 per cent because the market has been adequately developed and requires a smaller incentive to maintain consumer interest. The fiscal measures are complemented by a public TV and radio campaign to make the public aware of the tax credit, and an information campaign for all market participants in the three key renewable energy sections. Training and information for installers has been provided to encourage them to promote the tax credit to their customers. This programme has both 'carrot' and 'guidance' components. Both installers and end consumers are targeted by the guidance components.

The Household Subsidy Programme in Norway offered investment support and education measures for the implementation of heat pumps, wood pellet stoves and automated temperature controls. The subsidy covered 20 per cent of all costs including materials and installation, as long as the technology was purchased from a certified dealer and the installation was done by a certified installer. Guidance was offered through a telephone hotline staffed with engineers who were able to give good technical information and advice. The programme was delivered primarily on the internet through advertisements, webpages and the online application. During the initial programme, it was also publicized by the media. This is an example of a programme with both 'carrot' and 'guidance' components.

10.19 Best Practice 19: Design application process to be efficient for both homeowners and programme administration

Programme phase(s)

- Programme design.
- Programme evaluation, if an evaluation indicates that a complicated application process is a problem.

Market deployment stage(s)

- All.

Barrier(s) addressed

- Previous programmes (or the evaluated programme) have/has been hampered by complicated application processes. Both programme costs and public acceptance of the programme are affected by complicated processes.

Description

Many support programmes involve a process in which the homeowner applies for financial support. Having an efficient application process increases acceptance of the programme by the homeowner and reduces costs to the administration, which in turn increases acceptance with responsible policymakers. Support programmes should aim for an efficient process and to improve the process through review and feedback mechanisms.

Additional guidance

The following are recommended:

Aim for 'one-stop-shopping'

For the applicant, it is most efficient if they only have to apply once and only at one agency. In many cases this can also reduce the administrative costs and overheads. For example, by integrating the financial support for renewable heating installations into the building permit process, the homeowner may receive support without having to fill in additional forms or visit different government authorities. Often a check of the energy performance is already required before a building permit is granted. Integrating an eligibility check for a renewable heating installation into the process can be a very small additional administrative process, with minimal impact on the authority and almost unnoticeable to the homeowner. In the case of a tax rebate on the income tax, the application process can be integrated into the usual income tax declaration and review process.

Where a separate application process is unavoidable, for example in the case of a grant scheme for existing buildings, a one-phase process should be considered, where the applicant applies after the installation is done. This removes a very important barrier: many heating systems are installed at short notice because the existing one fails. In these situations, the homeowner often does not have the time to apply for support first. By being able to apply after the installation, the programme can give support in such urgent situations.

In cases where regions or municipalities offer financial incentives in addition to support given by the national government, the application process should not double the administrative effort. Where reasonable, the municipality could simply grant the support on the basis that national support has been granted.

Analysing costs and benefits of more complex application processes

The application procedure is typically used to assess whether a certain installation is or would be eligible for support. The process should ensure that the applicant, the product and possibly the planning/installation conform to certain criteria. Additional checks may increase compliance with the criteria and avoid fraud. However, it is important to balance the additional certainty with the negative consequences of a more complex application process. Where the consequences of a deviation from the parameters would be serious, a specific check in the application process is warranted. However, where the consequence would be small or where only a very small number of applicants would be involved, the increased complexity may not be reasonable. Costs and benefits of a further check and more complex application review should be well-balanced.

This best practice applies to all 'carrot' programmes which involve an application process. This is the case for most financial incentive programmes (except, for example, VAT reductions). It is also the case for 'stick' renewable obligations for new or refurbished buildings, when the house owner applies for a building permit.

Examples

The *Marktanreizprogramm* (Market Stimulation Programme or MAP) is the main instrument of the German federal government to stimulate the uptake of renewable heating and cooling technologies. In its latest edition, the MAP financially supports the installation of solar thermal, biomass heating and heat pump systems. The MAP has been running since 1999 and the grant approval process has been improved and stream-lined several times during this period. An important step towards reducing the time to process applications was introduced when the financial incentive was based on a rounded-up solar thermal collector area. Previously, preliminary grant approvals had been issued on the basis of the exact size of the collectors. However, this led to many recalculations after installation because the applied-for collector area differed slightly from the actually installed one. By granting a specific financial incentive for every rounded-up square metre of collector area, the need for these recalculations was drastically reduced. An even more important step, however, came in 2007, when the government decided to move to a one-step application process. Applicants now send their documents to administrators (BAFA) only after the installation, thus practically halving the effort and time needed to process one grant application. This simplification introduced some uncertainty as to the availability of budget: the homeowner who installs a renewable heating system cannot be sure that she or he will receive a grant, because the budget may be depleted by the time the installation is completed. To minimize this risk, BAFA publishes a 'Grant Signal' (*Förderampel*) on its website, which shows a green, orange or red light indicating the level of remaining budget. The experience with this tool so far is positive and the consumers have not been deterred by the higher risk.

The French tax credit programme was implemented as a direct fiscal measure for the support of renewable heating. The tax credit applies to household solar domestic hot water and other solar heating, biomass heating and heat pumps. From the beginning of the programme, a pre-approval process was eliminated by including the 'application' in the standard income tax declaration. This decision reduced costly administrative overhead and simplified the procedure for the consumer. By legally entitling the homeowner to the credit, the French programme also avoided any uncertainty caused by a possible lack of funds at the end of the fiscal year.

The energy strategy for the residential sector in Vorarlberg (Austria) includes two types of financial incentives provided to households for the installation of REHC technologies: the housing programme and a direct subsidy programme. Both are designed and financed by the state. The housing programme provides low-interest loans for building or renovating houses and the installation of district heating (biomass derived) or highly efficient gas, biomass or heat pump technologies. Except for in homes heated by district heating systems, the main heating system has to be combined with a solar thermal energy system. The Vorarlberg Regional Subsidy Programme provides additional direct subsidies for heat pumps, biomass heating and solar thermal installations with very strict technical requirements. Nevertheless, the application procedure has been kept very simple: the applicant provides an energy certificate for the building has the installation done and then turns in the application together with relevant invoices and the installer's affirmation of appropriate installation. Depending on the technology, additional documents concerning technical features can be required. The programme authority carries out random inspections of the installed systems to verify the provided information.

10.20 Best Practice 20: Ensure that the market infrastructure can successfully deliver and install products being promoted

Programme phase(s)

- Programme design.

Market deployment stage(s)

- All.

Barrier(s) addressed

- Market infrastructure is not well-established.

Description

To be successful, a support programme must take into account the local state of the market infrastructure. The local supply side is critical to actually putting the renewable heating and cooling systems in place and to supplying the necessary fuels (for biomass technologies). A support programme can help create the demand which is a precondition for the growth of a market infrastructure. It must not, however, be two steps ahead of the market, creating unrealistic expectations by the consumers and/or suppliers.

Additional guidance

This best practice can be implemented in two steps:

Analysis of the current state of the infrastructure in a specific market or market segment

The general analysis of market barriers/needs should help define which market segments to target next in order to further grow the market. It should also provide the information on the state of the market infrastructure. Are there enough companies offering the products and services that are to be supported? If not, could companies from neighbouring municipalities or regions help to set up the necessary businesses locally? Are personnel trained to provide the services needed? Are there administrative or regulatory issues that are holding back the market? Most of these questions are best discussed with the relevant stakeholder groups. The latter can often tell how easy or difficult it is to receive the services or products needed.

Identification of next steps necessary to develop the market

In cooperation with the representatives of the supply side (installers, manufacturers and so on), the programme authority should develop a plan for the development of the market or market segment in question.

This best practice applies to all technologies and is most relevant in early stages of a specific market or market segment, when the infrastructure to supply products and services may not yet be in place. It is important that the infrastructure of the specifically targeted market segment is analysed: for example, there may be a flourishing market for solar domestic hot water systems for one and two-family houses, but there may still be a lack of installers for solar combi+ systems (systems which supply domestic hot water and assist space heating and cooling).

Examples

The French tax credit programme was implemented as a direct fiscal measure for the support of renewable heating. The tax credit applies to household solar domestic hot water and other solar heating, biomass heating, and heat pumps. The programme was started in 2001 with relatively low financial incentives (15 per cent of the capital cost with a cap of €3007 for a single person). The programme launch coincided with the *Plan Soleil*, the national Solar Plan, which included other measures such as awareness-raising and training of professionals. This led to a slow but sustained growth of the solar thermal market, in which a healthy infrastructure could develop. The number of companies active in the French solar thermal increased significantly. The QualiSol qualification scheme for installers helped build up a network of qualified installers, which today includes more than 10,000 installers throughout France. After the initial phase, which lasted until 2004, the government stepped up financial support to 40 per cent and 50 per cent of the capital costs. Because the market infrastructure had started to evolve during the first phase, and because the programme focused on rather mature technology, the effect of the second phase was much stronger, increasing the solar thermal market by 80–120 per cent per year.

The *Spar mit Solar* programme (Steiermark, Austria) operated within the framework of the Austrian national promotion campaign *klima:aktiv*. Because the market for solar thermal installations had stagnated in 2003/2004 in Steiermark, this programme was designed to stimulate further market growth by providing general information on both the technology and the existing (regional) financial incentives provided by the state. The campaign cooperated closely with solar thermal dealers and installers to ensure that the numerous information events could be used afterwards for business contacts and lead to actual installations soon afterwards. Holding information events in areas with only low interest from solar thermal companies is not likely to yield good success.

The Barcelona Solar Thermal Ordinance, implemented by the municipality of Barcelona, requires that new buildings or buildings undergoing major renovations must have a solar domestic hot water system that meets at least 60 per cent of hot water demand. The ordinance was introduced first only for larger residential buildings. This effectively phased-in this new instrument and allowed for a subsequent built-up of the necessary market infrastructure. In 2006, the minimum energy demand was removed and today practically all new and refurbished buildings are covered by the obligation to install solar thermal systems.

The Canadian Residential Pilot Initiative for Solar Water Heating, part of the ecoEnergy for Renewable Heat programme, provides a slightly different lesson on the issue of market readiness. After the initiative was launched, programme administrators found that the existing codes and standards were not up-to-date for current solar technologies and were actually preventing solar technology deployment. However, they found that the offer of large government incentives led to high market demand, which put pressure on municipal building and plumbing inspection departments and the national regulatory body to urgently address the necessary revisions to the codes and standards. The administrators felt that, absent the incentives, progress on codes and standards revisions would have taken much longer.[8]

10.21 Best Practice 21: Design programme with adequate flexibility to adapt to market changes

10.21

Programme phase(s)

- Programme design.

Market deployment stage(s)

- All.

Barrier(s) addressed

- Market changes are anticipated during the course of the programme.

Description

The markets for renewable heating technologies are still characterized by a high level of innovation and rapid development. For long-term programmes, significant market changes are likely during the course of the programme. These changes can affect prices of both renewable and conventional heating systems and fuels. The technical state-of-the-art can also change, leading to improvements in efficiencies or emission outputs. Furthermore, policy and legal frameworks can change during the programme lifetime. Implementation or adjustment of relevant support programmes in the region or changes in housing regulations or legal emission thresholds may also have to be considered. (See also *Best Practice 29: Design programme to complement, not conflict, with other programmes in region*.)

Flexible programme designs can be ensured by planning regular programme revision milestones. Programme revision should be based on programme evaluation. (See also *Best Practice 28: Include an evaluation component in the programme*.)

Additional guidance

Both programme schedules and budget plans should be planned to provide flexibility. During programme design, regular revision milestones should be included in programme schedules and related costs should be considered. Revision milestones should be coupled with general or budgetary evaluations and revisions. A general market and technology review should be performed in order to identify changed or new needs of the markets addressed.

Examples

The *Spar mit Solar* programme (Steiermark, Austria) operated within the framework of the Austrian national promotion campaign *klima:aktiv*. Because the market for solar thermal installations had stagnated in 2003/2004 in Steiermark, this programme was designed to initiate further market growth by overcoming the existing lack of general information on both the technology and the existing (regional) financial incentives provided by the state. At the beginning, the campaign targeted builders and owners of detached houses because they represented the largest share of the residential market. After a solar ordinance was enacted and enforced, solar thermal installations in new homes became mandatory when applying for low-interest housing loans. The focus of the campaign was therefore shifted to the retrofit sector. The content of events was adjusted and, more importantly, the promotion strategy, including the slogan, was adjusted to target the retrofit sector.

In Germany, a renewable heating law has come into effect requiring the use of renewable energy to meet a specified portion of building loads. The very successful MAP has adapted by lowering the incentives available for new buildings and restricting those incentives to installations providing more than the required amount of energy.

The French tax credit programme was implemented as a direct fiscal measure for the support of renewable heating. The tax credit applies to household solar domestic hot water and other solar heating, biomass heating and heat pumps. The programme has been adjusted several times: the initial credit, implemented from 2001–2004, was for 15 per cent of the capital cost of the equipment. The uptake of the programme was low so that the second time period of the credit (2005–2009) allowed a 40 per cent credit, which was raised again to 50 per cent (as of January 2006). For 2009, the credit for biomass heating was reduced to 40 per cent because the market had been adequately developed and required a smaller incentive to maintain consumer interest.

The Solar Keymark programme provides a contrast to the above examples. This programme is a certification mark owned by the European Committee for Standardization (CEN). It certifies conformity of a product with the applicable European Standards, EN 12975 for solar thermal collectors or EN 12976 for factory-made solar thermal systems. The Solar Keymark is the solar thermal specific edition of the general Keymark certification scheme. Due to a lack of flexibility of the general Keymark system and certification rules, it is difficult to extend and update the standards and scheme rules to the latest solar technological and market developments. For example, today there is no European Standard covering air collectors, and therefore no Solar Keymark can be issued for this technology

10.22 Best Practice 22: Engage media to provide publicity

Programme phase(s)

- Programme design.
- Programme implementation.

Market deployment stage(s)

- All.

Barrier(s) addressed

- Public relations budget for programme is too small or the public is not adequately informed about the programme for other reasons.

Description

Publicity is critical to the success of several programme types. Subsidy programmes or other financial incentive instruments can only achieve a large impact when potential users are aware of them. Guidance programmes, such as information campaigns, depend largely on publicity. Solar information campaigns in Austria demonstrated that event coverage and event announcement in regional media was important for public awareness-raising and consequently for the impact of such campaigns. Establishing long-term media contacts and frequently providing information, particularly at the beginning of the programme, is one method to reach broad target groups.

Additional guidance

The identification of suitable media partners and first steps towards establishing media partnerships should be part of the design process. During the implementation phase, media partners should be continuously supplied with information, both general promotion materials (brochures, and so on) and information on past and future events.

This best practice applies to all programmes for which public awareness is critical. Opportunities provided by financial incentive programmes must be known to potential users in order to influence decisions on heating systems. However, in some cases, installers inform their customers. In these cases it is also feasible to target installers with information campaigns. This can be easily done by providing the necessary information to professional organizations and industry-specific media such as trade publications.

Examples

The *Spar mit Solar* programme (Steiermark, Austria) operated within the framework of the Austrian national promotion campaign *klima:aktiv*. Because the market for solar thermal installations had stagnated in 2003/2004 in Steiermark, this programme was designed to initiate further market growth by providing general information on both the technology and the existing regional financial incentives provided by the state. One problem encountered at the beginning of the *Spar mit Solar* promotion programme was that the interest of the media in general was not sufficient. This was overcome by establishing continuous contacts with the media and by repeatedly providing information to both print media and TV. This increased media interest significantly over the course of the project. The media strategy included cooperation with a regional print medium group which is represented within the region by a number of subregional daughter media. The advertising of events and the coverage of the events were done by these subregional newspapers.

10.23 Best Practice 23: Develop and distribute tools that facilitate household participation

Programme phase(s)

- Programme design.
- Programme implementation.

Market deployment stage(s)

- All.

Barrier(s) addressed

- Developing desired programme participation is expected to be difficult or programme participation is not as high as expected.

Description

Uncertainties about the technical and economical viability of renewable heating technologies are one of the major barriers towards further technology uptake. Tools that provide basic information on new technologies to end-consumers are one method of reducing these uncertainties. For example, the decision on heating systems in households can be facilitated by providing tools such as heating cost calculators that allow for an objective comparison of different heat options.

As soon as the end-consumer has decided to invest in renewable heating, he faces other difficulties. For example, it can be difficult for the end-consumer to find and choose an adequate installer. Online databases listing installers, technology providers or financing opportunities (such as available subsidies) have been shown to be useful tools to facilitate household participation in renewable heating markets.

Additional guidance

Several best practices identified in this report will be most easily implemented after a market study is conducted. This study should include collecting the following information:

- **Market status:** Who are the manufacturers? Wholesalers? Retailers? Installers? Operations and maintenance providers? End consumers? Can the existing market infrastructure respond quickly to an increase in demand? (For energy-efficient biomass technologies, both the equipment and fuel supply chain should be included in the market status study.)
- **Market needs and barriers:** What are the needs and barriers, both perceived and real, identified by all of the market players?
- **Existing programmes:** What existing programmes are operating in the region? How are these programmes addressing the identified needs and barriers?

During the course of this market study, a market barrier or need may be identified that can be addressed by the creation and distribution of a tool.

Examples

Several public and private organizations provide heating cost calculators aimed at facilitating the end-consumers' decisions about renewable heating. A very simple tool distributed by Coed Cymru in Wales, UK,[9] compares the cost of wood chips and pellets to conventional fuels. The EU-financed Biohousing project provides a life cycle cost calculator for various home-heating technologies.[10]

Assistance in finding suitable installers can be provided by listing certified installers in public databases. In France, *Qualit'EnR*[11] organizes the certification of installers of solar thermal technologies (*Qualisol*), biomass heating (*Qualibois*) and heat pumps (*Qualipac*).

Tools for simplifying the bidding process also facilitate consumer decisions. Standardized bidding documents were distributed during the Austrian *Ja zu Solar* programme.

The online information platform by Sustainable Energy Ireland[12] provides a good example of complete consumer information. The website not only contains general information and guides on residential renewable heating but also provides databases of qualified installers, equipment and available funding schemes.

Databases on available state incentives are provided, among others, by the US Department of Energy[13] (general US) and by proPellets Austria[14] (specifically on pellet incentives in Austria).

10.24 Best Practice 24: Invest in high-quality promotion materials for information campaigns

Programme phase(s)

- Programme design.

Market deployment stage(s)

- All.

Barrier(s) addressed

- Public and/or industry is not well-informed about technology or application.
- Programme is such that public and/or industry may get confused about technical and programme details.

Description

The aim of a renewable energy image or information campaign is, in addition to explaining the obvious environmental benefits, to show the reliability and cost-efficiency of renewable heating technologies to the end-consumer. Promotion materials must communicate clear and simple messages;

end-consumers should not be confused by unnecessary details. Moreover, the appropriate promotion materials may change as the programme progresses: general information is likely to be required during an initial awareness-building phase of the project, with more specific information available later when the programme is in progress. The overall quality and design should be attractive enough to capture the reader's attention and convey an impression of professionalism and reliability. To meet these needs, investing in professional development of promotion materials may be required.

Additional guidance

The communication skills and interests of the administrative organization should be evaluated. If this organization does not have the skills or interest in developing high-quality promotional materials, a third party, such as a professional communication agency, should be engaged. The programme budget must be adjusted accordingly.

Examples

The local regional energy agency in Tirol (*Energie Tirol*, Austria) started the *Ja zu Solar* information campaign in 2004. Since a lack of information on the consumer side was identified as one of the main barriers to increased uptake of solar thermal technologies, the implementation of a broad information campaign was seen as an appropriate tool to reach this goal. A professional communication agency was contracted to design promotion materials and the slogan.

The *Spar mit Solar* (Steiermark, Austria) programme operated within the framework of the Austrian national promotion campaign *klima:aktiv*. The campaign was initiated because the government had established that a lack of information on the part of the consumers was the main barrier for the broad deployment of solar thermal technology in Steiermark. A professional media design partner provided promotion materials such as flyers, brochures, slogans and logos.

10.24

10.25

10.25 Best Practice 25: Ensure a central point for accurate information

Programme phase(s)

- Programme design.
- Programme implementation.

Market deployment stage(s)

- All.

Barrier(s) addressed

- Programme is such that public and/or industry may get confused about technical and programme details.

Description

Uncertainties created by inconsistent information provided by different organizations can negatively affect programme success. The implementation of central contact points, including the provision of advice via telephone hotlines, has shown to be one method of avoiding confusion. Regularly updated programme descriptions presenting comprehensive information should be also be provided by one central organization.

Additional guidance

One central contact point for providing information to consumers, retailers and installers should be established in order to prevent the dissemination of inconsistent information. The contact point can be provided by the organization responsible for programme administration or by an external partner such as an energy agency. Part of the strategy should be to produce uniform dissemination materials such as short brochures and comprehensive programme descriptions answering the majority of possible questions. It is crucial that these materials are adapted quickly to changing programme directives and it must be made clear which versions are valid at a given time. Additional questions can be handled by telephone hotlines.

Examples

The Household Subsidy Programme in Norway offered investment support and education measures for the implementation of heat pumps, wood pellet stoves and automated temperature controls. The subsidy covered 20 per cent of all costs including materials and installation, as long as the technology was purchased from a certified dealer and the installation was done by a certified installer. Enova SF, the public agency running the programme, provided consistent information on the programme to the public and served as a central contact point.

The French tax credit programme provides a counter example. This programme was implemented as a direct fiscal measure for the support of renewable heating. The tax credit applies to solar heating, biomass heating and heat pumps. One challenge for the programme was that customers were given different information by installers, manufacturers and tax officials. In some cases, customers were not clear on what equipment was eligible for incentive. For heat pumps, for example, only the outdoor equipment may be included in the tax credit. A central information point might strengthen this programme.

10.26 Best Practice 26: Design programme with entry and exit strategies

Programme phase(s)

- Programme design.

Market deployment stage(s)

- All.

Barrier(s) addressed

- Programme will need to terminate at some point.

Description

Information from several programmes indicates that market changes occur when programmes start, end or change. This may be beneficial if the result is positive (market prices for a technology decrease, for example). However, it may be harmful if market distortions occur in order to maximize incentives. To prevent undesirable market distortions, programme design should include entry and exit strategies. For long-term programmes, possible market distortions caused by periodic programme adjustments should be considered.

Additional guidance

Entry and exit strategies must be considered during programme design. One strategy to avoid negative impacts due to programme termination, depleted budgets or programme adjustments might be to announce relevant information at short notice. A strategy for moving a maturing market away from dependence on incentive programmes might be to systematically decrease the incentives over time.

This best practice applies to a wide range of programmes independent of technology and is particularly important when budgets are fixed for certain periods. Under these circumstances, strategies for dealing with depleted funds must be developed.

Examples

The Household Subsidy Programme in Norway offered investment support and education measures for the implementation of heat pumps, wood pellet stoves and automated temperature controls. The heat pumps subsidized in the programme included air-to-air heat pumps, which are the least expensive heat pump technology. After the programme, it was determined that a subsidy was no longer needed for this technology because the price in the Norwegian market had become competitive with traditional heating technologies. Consequently, in the 2006 programme, the air-to-air heat pump was not subsidized, although air-to-liquid heat pumps for in-floor heating were. In this case, the market change was positive and mitigating entry or exit strategies were not required.

The *Marktanreizprogramm* (Market Stimulation Programme or MAP) is the main instrument of the German federal government to stimulate the uptake of renewable heating and cooling technologies. In its latest edition, the MAP financially supports the installation of solar thermal systems, biomass heating installations and heat pump systems. The MAP is financed from the public budget and therefore needs annual approval. This annual approval has created uncertainty in the market, disrupting the otherwise strong development of the market for renewable heating and cooling. The depletion of the annual budget contributed to the significant decreases of the solar thermal market in 2002 and 2007. The German government now announces changes with lead times of just a few days. In 2006, when the MAP budget was depleted by autumn, the government took the unusual step of announcing that in the first months of 2007, applications could be filed for those systems installed in 2006, when no budget was available. This helped prevent a drastic downturn otherwise expected for the end of the year. (More recently, the 2008 Renewable Heating Law was intended to provide more stability at lower cost to the public budget.) Experience with MAP and with financial incentive schemes in other countries showed that the discussion of changes to a grant scheme has strong and negative effects on the market stability. In cases where a decrease of the grants is discussed, consumers rush to still receive the (higher) grants, leading to a temporary overheating of the market and then a phase of sharp decline. When an increase in grants is discussed, consumers typically postpone their investment decision until after the new regulations come into effect. This can bring the market to an almost complete halt until after the new grants are in place.

10.27 Best Practice 27: Get committed renewable energy champions to participate

Programme phase(s)

- Programme design.
- Programme implementation.

Market deployment stage(s)

- All.

Barrier(s) addressed

- Legal, administrative or funding barriers will hamper the programme.
- Public is not adequately informed about programme.

Description

The close cooperation of programme administrators and renewable energy champions is important for the success of support programmes. Public figures who support renewable energy can provide publicity for programmes and technologies. More specifically, politicians who favour renewable energy can assist not only with publicity but also in removing legal or administrative barriers that might interfere with programmes and in lobbying for adequate programme budgets.

Additional guidance

A strategy to convince champions to participate might be to provide positive news derived from programme activities that can be communicated to the public by these champions.

Examples

The Barcelona Solar Thermal Ordinance, implemented by the municipality of Barcelona, requires that new buildings or buildings undergoing major renovations have a solar domestic hot water system that meets at least 60 per cent of hot water demand. Municipal politicians, in collaboration with non-governmental organizations and solar companies, were involved in the development of the ordinance.

10.28 Best Practice 28: Include an evaluation component in the programme design

Programme phase(s)

- Programme design.

Market deployment stage(s)

- All.

Barrier(s) addressed

- Programme concept has not been tested in target population. (Implementing an evaluation component is especially useful when a new type of programme is implemented or when a programme is implemented under specific regional or market circumstances.)
- Long-term programme will need periodic adjustments.

Description

Many of the successful programmes analysed during this project are long-term programmes. The confidence created by longer programme durations provides a strong driver for renewable heating market uptake (see *Best Practice 11: Design and implement long-term programmes*). Since the markets for renewable heating technologies are still characterized by a high level of innovation and rapid development, significant market changes during long-term programme implementation are likely. The adjustment of long-term programmes to changing technical, political and economic conditions therefore is critical to programme success (see *Best Practice 21: Design programmes with adequate flexibility*). Programme flexibility can be ensured by planning regular programme revision milestones. Programme revision should be based on thorough programme evaluation.

Evaluation can include: the comparison of programme impact against set targets; the satisfaction of applicants and other involved stakeholders; the appropriateness of programme features, such as offered subsidy amounts; and the cost and time effectiveness of programme administration. The information provided by programme evaluation is not only useful for adjusting existing programmes, but also for informing the design of future programmes.

Additional guidance

Revision milestones should be scheduled during programme design. If programme budgets are revised annually, general programme revision can be done in parallel. Evaluation information can be used to prepare necessary programme adjustments and to improve budget decisions. In many successful programmes, evaluation is provided by external expert organizations. These can be energy agencies, consultants or research institutes. This encourages cost-efficient and independent programme assessment. Topics addressed should include: the comparison of programme impact against set targets; the satisfaction of applicants and other involved stakeholders; the appropriateness of programme features, such as offered subsidy amounts; and the cost and time effectiveness of programme administration.

Examples

The Household Subsidy Programme in Norway offered economic support and education measures for the implementation of heat pumps, wood pellet stoves and automated temperature controls. A social cost/benefit analysis was completed for the programme. This evaluation took into account the social benefits such as saved energy costs, reduced grid power challenges and reduced market prices for the products. Programme administration costs, technology and pellet fuel costs were also evaluated and a household survey was conducted. This evaluation of the 2003 programme was used to inform the 2006 programme.

The energy strategy for the residential sector in Vorarlberg (Austria) includes two types of financial incentives provided to households for the installation of

REHC technologies: the housing programme and a direct subsidy programme. Both are designed and financed by the state. The housing programme provides low-interest loans for building or renovating houses and the installation of district heating (biomass derived) or highly efficient gas, biomass or heat pump technologies. Except for homes on district heating systems, the main heating system has to be combined with a solar thermal energy system. The state of Vorarlberg provides additional direct subsidies for heat pumps, biomass heating and solar thermal installations. The direct subsidy programme performance indicators are the collector area installed and the number of installations compared to the resulting programme costs. *Energieinstitut Vorarlberg* provides in-depth analyses of programme statistics. Furthermore, measurement programmes (for example, measuring the efficiency of heat pumps) and on-site inspections of approved installations are done. This data indicates for which sector or technology better support is needed and allows for the identification of quality problems. The results can then be used for the regular amendments of the programme directives.

The state of Steiermark provides direct subsidies for the installation of efficient biomass heating and solar thermal collectors through the *Umweltlandesfonds*. The *Umweltlandesfonds* provides reports on its activities annually to the government. In addition, the *Umweltlandesfonds'* work from 2002–2007 was thoroughly assessed by the general accounting office (*Landesrechnungshof*) in 2007. The budget-handling, the design of the directives and the organization of the application-handling were assessed and detailed recommendations were provided.

The *Marktanreizprogramm* (Market Stimulation Programme or MAP) is the main instrument of the German federal government to stimulate the uptake of renewable heating and cooling technologies. In its latest edition, the MAP financially supports the installation of solar thermal systems, biomass heating installations and heat pump systems. The MAP has been regularly evaluated by several independent scientific institutes (ZSW, Solites, ITW and others). Detailed evaluation reports exist for 2002–2004, 2004–2005 and 2006. A short version for the first year of the two-year period 2007–2008 was published in 2008. Topics addressed in these reports are the impact of the programme on technology market penetration, the relation between the financial incentive and total installation costs, development of, and programme impact on, technology prices and the contribution of the MAP to greenhouse gas emission reduction. The studies are based on the available data from the Federal Office for Economy and Export Control (BAFA) and the public KfW bank and have frequently been complemented by interviews/questionnaires of market participants (planners, installers, manufacturers, industry associations and end-users). The evaluation reports not only study the workings of the previous period but also give recommendations for the further improvement of the programme. Many of the recommendations, including the one to move towards a one-step application process, were later taken up by the government.

10.29 Best Practice 29: Design programme to complement, not conflict, with other programmes in region

Programme phase(s)

- Programme design.
- Programme implementation or evaluation if market conditions change.

Market deployment stage(s)

- All.

Barrier(s) addressed

- Other REHC or energy conservation programmes are operating in the target region.

Description

When other programmes are available in the same geographical region, new programmes should be designed to complement, not conflict, with the existing programmes. For example, if a subsidy programme for solar hot water is already available from another organization or level of government, a new programme could complement the existing programme by targeting a niche group not covered by the existing programme, providing an add-on subsidy to further decrease homeowner costs or running an educational campaign to inform the public about the benefits of solar energy (and the subsidies available).

Additional guidance

Several best practices identified in this report will be most easily implemented after a market study is conducted. This study should include collecting the following information:

- **Market status:** Who are the manufacturers? Wholesalers? Retailers? Installers? Operations and maintenance providers? End consumers? Can the existing market infrastructure respond quickly to an increase in demand? (For energy-efficient biomass technologies, both the equipment and fuel supply chain should be included in the market status study.)
- **Market needs and barriers:** What are the needs and barriers, both perceived and real, identified by all of the market players?

- **Existing programmes:** What existing programmes are operating in the region? How are these programmes addressing the identified needs and barriers?

The information gathered during this study should allow programme designers to identify currently unaddressed needs and barriers which can then be targeted by a complimentary new programme.

Although this best practice applies to all programmes, it will be most important for programmes developed and implemented at the municipal or regional level, which will need to avoid conflict with programmes covering a wider geographical area. Programmes developed and implemented at the national or even continent-wide level may not be able to avoid conflicts with smaller-scale existing programmes.

Examples

Five regional Austrian programmes were studied during this project. Two are education programmes: *Spar mit Solar* and *Ja zu Solar*. Three are subsidy ('carrot') programmes: Salzburg Regional Subsidy Programme, Vorarlberg Regional Subsidy Programme and the Steiermark Subsidy Programme. All of these programmes complement the Austrian national renewable campaign *klima:aktiv* and were, in some cases, designed with assistance from *klima:aktiv*. The three regional subsidy programmes studied target niche markets that are not eligible for the regional home loan plans that subsidize renewable energy installations in new or renovated homes.

During the programme identification phase of this project, we also identified programmes that were less successful because they competed directly with other programmes. For example, during the first period (2001–2004) of the French direct tax credit programme, the solar and biomass tax credits competed directly with subsidies available from Agence de l'Environnement et de la Maîtrise de l'Energie (ADEME). This may account for the low number of households taking advantage of the tax credit during this period. However, the tax credit was increased from 15 per cent to 40–50 per cent of the equipment capital cost for the second programme period (2005–2009) and the ADEME subsidies were changed to target energy efficiency and support complementary regional programmes. The incentives no longer competed with each other, leading to a significant increase in the number of households taking advantage of the credit.[15]

Even successful existing programmes must adapt when new programmes in the same region present conflicts. For example, a renewable heating law is now in effect in Germany, requiring the use of renewable energy to meet a specified portion of building loads. The very successful MAP has adapted by lowering the incentives available for new buildings and restricting those incentives to installations providing more than the required amount of energy.

10.29

Finally, care should be taken that the new subsidy does not render the existing subsidy obsolete or ineffective by comparison. For example, in Canada in 2009, the ecoEnergy Retrofit programme established a retrofit solar water heating subsidy far more lucrative than the existing new home subsidy. As a result, new homeowners waited to install their solar system until after their premises had been lived in for six months, thereby qualifying them for the retrofit subsidy.[16]

10.30 Best Practice 30: Limit information campaigns to specific, clear messages

Programme phase(s)

- Programme implementation.

Market deployment stage(s)

- All.

Barrier(s) addressed

- Public is not adequately informed about the programme.

Description

Guidance programmes are most effective when they address only one technology or a limited set of technologies or even specific, well-identified market barriers. This allows conveying the most important messages without confusing the end-consumer.

Additional guidance

Several best practices identified in this report will be most easily implemented after a market study is conducted. This study should include collecting the following information:

- **Market status:** Who are the manufacturers? Wholesalers? Retailers? Installers? Operations and maintenance providers? End-consumers? Can the existing market infrastructure respond quickly to an increase in demand? (For energy-efficient biomass technologies, both the equipment and fuel supply chain should be included in the market status study.)
- **Market needs and barriers:** What are the needs and barriers, both perceived and real, identified by all of the market players?
- **Existing programmes:** What existing programmes are operating in the region? How are these programmes addressing the identified needs and barriers?

The information gathered during this study should allow programme designers to identify specific market barriers. Appropriate information campaign messages can then be formulated to address these market barriers. Appropriate delivery mechanisms can also then be identified. In cases where more than one type of market participant (for example, end-consumer, installer, retailer) needs to be targeted, the content and the delivery mechanism may be different for each type of participant.

Examples

The *Spar mit Solar* programme (Steiermark, Austria) operated within the framework of the Austrian national promotion campaign *klima:aktiv*. During the programme, eight to ten information events were organized per year. A general lack of information on solar thermal technologies and a lack of trust in the technological and economical viability of solar thermal technologies were identified as the main barriers to further growth of the market. The events were not only exclusively addressing solar thermal technologies but specifically highlighting the technological maturity of solar thermal technologies and providing information on equipment prices, payback times and ways to decrease investment costs (such as participating in the regional subsidy programme).

10.31 Best Practice 31: Keep business interests in the background during information campaigns

Programme phase(s)

- Programme implementation.

Market deployment stage(s)

- All.

Barrier(s) addressed

- Participation by industry representatives may be seen as a conflict of interest.

Description

Image or information campaigns can profit directly from the involvement of industrial partners, such as technology manufacturers and installers, who are willing to contribute either time or funding to guidance programmes. Indirectly, their involvement can increase the impact of campaigns by facilitating contact between business and potential customers. This can lead to additional installations.

   On the other hand, information campaigns typically aim to reduce potential end-consumers' mistrust of new technologies and the business interests of involved industries. Therefore, the information should be presented by independent experts (for example, from semi-public bodies such as energy agencies) while business interests are kept in the background. It is essential that information events do not appear to be mere sales events.

Additional guidance

According to the Austrian experience, it is feasible to organize small-scale exhibitions after information events, where industry partners can present themselves and get in contact with potential customers.

Examples

The local regional energy agency in Tirol (*Energie Tirol,* Austria) started the *Ja zu Solar* information campaign in 2004. Since a lack of information on the consumer side was identified as one of the main barriers to increased uptake of solar thermal technologies, the implementation of a broad information campaign was seen as an appropriate tool to reach this goal. The financing industrial partners were allowed to present their companies after the events in order to reach potential customers. They were kept in the background during the information event and their presentation was strictly limited to flyers and a poster. It was seen as very important to organize information events, not sales events.

The *Spar mit Solar* programme (Steiermark, Austria) operated within the framework of the Austrian national promotion campaign *klima:aktiv*. Because the market for solar thermal installations had stagnated in 2003/2004 in Steiermark, this programme was designed to stimulate further market growth by presenting information on both the technology and the existing regional financial incentives provided by the state. During the information events, only independent experts were in contact with visitors, while the industrial partners stayed in the background. After the event, the industrial sponsors could come in contact with potential customers by presenting their companies in an attached exhibition.

10.32 Best Practice 32: Monitor and publicize progress towards targets

Programme phase(s)

- Programme implementation.

Market deployment stage(s)

- All.

Barrier(s) addressed

- Lack of motivation with long-term objectives because it is difficult to tell what progress is being made.

Description

Monitoring and reporting of progress is needed to demonstrate accountability and to provide information for corrective action to improve policies and programmes. Programme evaluations provide periodic comprehensive information (usually at intervals of several years), but routine, ongoing performance reporting is also needed, both at the level of programmes and at the broader portfolio level.

Additional guidance

The establishment of milestone or interim targets should be accompanied by the development of an appropriate measurement, verification and reporting system, using accepted standards for measurement and verification. Mechanisms should be established to receive reports and take corrective action if necessary (for example, by increasing incentives or raising the price of conventional fuels) to get progress on track to achieve the next interim target.

Monitoring progress through measurement and verification may be of particular importance when predicted equipment performance, and therefore support programme accomplishments, have been based on computer simulations. Actual measurements of installed systems may be helpful for encouraging participation from both the public and industry.

Example

The Climate Alliance of European Cities is a network of European cities and municipalities working at the municipal level on climate protection. These cities and municipalities have entered into a partnership with indigenous rain forest peoples of the Amazon. There are two main objectives of the Climate Alliance: to reduce greenhouse gas emissions to a sustainable level in the northern hemisphere and to support projects to conserve the rain forest and improve quality of life in the southern hemisphere through cooperative programmes.

With support from the 17 member municipalities and districts in Germany, the Climate Alliance has developed software to monitor savings. The software is the energy and CO_2 calculation tool ECO2-Regio from the Swiss company Ecospeed, modified to suit the Climate Alliance. So far the software is available only in Germany and Switzerland, with 100 local and regional organizations using the software. Each municipality must pay €350 to access the software. Currently, Climate Alliance Italy is working to develop a version for the Italian municipality members of the alliance.

10.32

Also, an annual activity report is published to assess the municipalities and programmes supported by the Climate Alliance. This report describes the success of some of the programmes involved and sets out plans for the next year.

Notes

1   This description is based on the authors' observations of typical approaches to energy efficiency and renewable energy policy over the last 20 years.
2   This description is based on the authors' observation and experience with effective frameworks in the context of energy efficiency and renewable energy over the past 20 years.
3   See http://portal.foerdermanager.net.
4   *Solar Water Heater (SWH) Owners Guide*, Australian Government Office of the Renewable Energy Regulator, www.orer.gov.au/swh, accessed September 2009.
5   'Directive 2002/91/EC of the European Parliament and of the Council of 16 December 2002 on the energy performance of buildings', *Official Journal L 001*, 04/01/2003 P. 0065–0071 in the EurLex database, http://eur-lex.europa.eu/LexUriServ/LexUriServ.do?uri=CELEX:32002L0091:EN:HTML, accessed July 2009.
6   *What is the EnerGuide Rating System?*, Natural Resources Canada, http://oee.nrcan.gc.ca/residential/personal/new-homes/upgrade-packages/energuide-service.cfm?attr=4, accessed July 2009.
7   *ecoENERGY Retrofit – Homes Program*, Natural Resources Canada, http://oee.nrcan.gc.ca/residential/personal/grants.cfm?attr=4, accessed July 2009.
8   Personal communication with Jeff Knapp of the Renewable and Electrical Energy Division of Natural Resources Canada.
9   *Heat Cost Calculator and Comparator*, Coed Cymru, www.coedcymru.org.uk/calculator.htm, accessed July 2009.
10   *Biohousing Heating Tool: Energy Calculator,* Biohousing website, www.biohousing.eu.com/heatingtool/ecalcmap.asp, accessed July 2009.
11   Qualit'EnR website, www.qualit-enr.org, accessed July 2009.
12   Sustainable Energy Ireland website, www.sei.ie, accessed July 2009.
13   *DSIRE: Database of State Incentives for Renewables and Efficiency*, www.dsireusa.org, accessed July 2009.
14   Propellets website, www.propellets.at, accessed July 2009.
15   Observ'ER, *REFUND+ Quantitative and Economic Assessment of the Direct Fiscal Measures: Study of the French Case*, Angers, France: ADEME, 2007. Available at http://ftpnrj.free.fr/refund+/French_economic_report.pdf.
16   Personal communication with Jeff Knapp of the Renewable and Electrical Energy Division of Natural Resources Canada.

11
Part 2 Glossary

Abbreviations

ADEME: Environment and Energy Management Agency (France)
BAFA: Federal Office for Economy and Export Control (Germany)
GDP (PPP): gross domestic product adjusted for purchasing power parity
GDP: gross domestic product
IEA: International Energy Agency
MAP: Market Incentive Programme (Germany)
OECD: Organisation for Economic Co-operation and Development
REHC: renewable energy heating and cooling

Terms

Achievable potential: An estimate of displaced consumption of energy from traditional sources that can be realistically achieved given institutional, economic and market barriers. Achievable potential is typically some fraction of economic potential.

Air-source heat pump: A heat pump which uses the outside air as a heat sink or source.

Air-to-air heat pump: A heat pump which uses air as both a heat source and heat sink.

Air-to-liquid heat pump: A heat pump system in which a liquid is circulated in the conditioned space to provide heating or cooling. Air is used as the heat source or sink.

Biomass: Vegetable matter used as a source of energy. Biomass materials examined in this study include pellets, wood and wood waste.

Barrier: Any condition that discourages the use of a given renewable technology. For example, high cost would be an affordability barrier. Lack of qualified installers would be a availability barrier.

Economic incentives: Grants, loans, rebates and/or tax credits used to encourage the adoption of renewable technologies. Also referred to as 'carrots'.

Economic potential: An estimate of displaced consumption of energy from traditional sources if all cost effective REHC technologies were implemented. Economic potential is typically some fraction of technical potential.

Full market stage: At the full market stage, the public is knowledgeable about the target technology and its benefits, and the technology can be easily obtained and installed.

Gross domestic product adjusted by purchasing power parity: An adjustment to GDP which takes into consideration the price of goods traded within a country (based on a specific 'basket of goods') rather than considering only the export value and exchange rate of the country's currency.

Gross domestic product: The total market value of all goods and services produced in a country in a given year.

Ground-source heat pumps: Heat pumps which uses the Earth as both a heat source and heat sink.

Guidance: See Information and marketing.

Heat pump: A device which uses mechanical energy to transfer heat from one place (the 'source') to another (the 'sink').

Information and marketing: A type of renewable energy promotion programme that includes public education and awareness campaigns, quality assurance standards, support resources and training. Also referred to as 'guidance'.

Initial deployment market stage: At the initial deployment market stage, clients may be largely unaware of the promoted technology, technology may not be readily available or installation and maintenance support may be difficult to obtain. Programme activities may include demonstration and pilot projects and training for retailers and installers.

Levelized unit energy cost: The life cycle cost of producing one unit of useful energy by means of a given technology. Used to compare the attractiveness of various energy supply options.

Mass market stage: At the mass market stage, target technologies are becoming known to the public and are available through retailers or installers. Programmes may include 'carrot' and 'guidance' activities.

Multi-unit dwellings: A residential building intended to house more than one family, such as an apartment building or row house.

Passive solar thermal systems: Solar thermal systems that do not involve moving components such as pumps. These are distinct from 'active solar thermal systems', which typically involve moving components.

Pellets: Pulverized biomass materials compressed into pellet form in order to ease transport and handling and combusted to provide heat.

Portfolio planning: Programme planning phase encompassing initial problem definition, setting objectives, identifying intervention points, identifying stakeholders and factors and selecting instruments.

Programme design: Programme phase preceding implementation, and ideally based on the portfolio planning phase. May be undertaken by the funding agency or a third party.

Programme evaluation: An assessment of an ongoing or completed programme against objectives. May include an assessment of problems encountered, relevance, implementation issues, and cost-effectiveness.

Programme implementation: Phase in which the programme is delivered to end-users. This phase may be undertaken by the funding agency or a third party.

Programmes: In this report, this term is used to refer to any organized effort to encourage the use of renewable heating and cooling technologies. It includes efforts that may be referred to elsewhere as 'policies'.

Regulations: Building codes and/or minimum standards requiring the use of renewable technologies. Also referred to as 'sticks'.

Sticks: See 'Regulations'.

Technical potential: An estimate of displaced consumption of energy from traditional sources if all technically viable REHC technologies were implemented immediately in every appropriate application, regardless of cost.

Index